Guidebook on Mouse and Rat Colony Management

Contributors and affiliations:

Kathleen R. Pritchett-Corning, DVM, DACLAM, MRCVS
Director, Research and Professional Services
Research Models and Services
Charles River

Sonja T. Chou, VMD, MS, DACLAM
Director, Veterinary Services
Preclinical Services
Charles River

Laura A. Conour, DVM, DACLAM
Director of Laboratory Animal Research
Princeton University

Bruce J. Elder, Ph.D.
Director, Corporate Rodent Genetics
Genetically Engineered Models and Services
Charles River

Published by Charles River Laboratories, October 2011

Table of contents

It is estimated that mice and rats comprise more than 80% of the animals used in research in the United States, with mice far outnumbering rats. In countries where detailed statistics are kept, such as the UK, in 2010, mice were involved in 72% of all animal scientific procedures, and rats 8%.[1] Why have rodents come to dominate science? There are many reasons, some of which are outlined below.

Mice are the model mammalian system for scientific enquiry. This follows from their initial use as the model mammalian genetics organism. Mice were the first mammals after humans to have their genome sequenced.[2] Sequencing followed over 80 years of empirical genetics research in mice, and at least 1000 years of practical genetics research by mouse fanciers from ancient China to today. As mammals, mice have more similarities to than differences from humans. There is approximately 85% homology between mouse and human genes, which means that a particular gene is most likely present in both the mouse and human and will generally have a similar function and protein product.[2] This allows mice to serve as models of many human conditions, and, more importantly, allows us to study basic mammalian genetics and other conserved systems in mammalian cells.

As research subjects, mice have other advantages as well. Compared to other laboratory animals, mice require relatively little space. They are easy to physically manipulate, can be gentled to human handling, and do not have complex dietary needs. Again, compared to other laboratory animals, they are relatively inexpensive, although individual genetically manipulated mice can be costly. As small mammals, they have a limited lifespan, which makes aging and multigenerational studies easier. Mice can be inbred, something many species do not tolerate well. Many mice that we work with today are as alike as possible without being clones, which can be useful in studies where variation is not desired.

The mouse's reproductive biology is another factor in their utility in research. Mice are polyestrous, litter-bearing animals, with a short gestation period. Since they are polyestrous, they will reproduce year-round and the fact that they are litter-bearing allows for the potential for littermate controls within each litter and quickly engenders many animals for scientific study. The short gestation period means that multiple generations can be produced rapidly and followed for experimental purposes. Given ideal conditions, mice can produce at least four generations in a year. Mice are also quick to reach sexual maturity and able to reproduce at 4-8 weeks of age. Their fertility with subspecies such as *Mus musculus castaneus* allows for genetic mapping and study of the influence of the surrounding genome on specific genes. A further explanation for the dominance of the mouse in research is the robustness of their embryos. These may be cryopreserved and cultured from one-cell to post-implantation stages. Tools have been developed that allow for sophisticated manipulations of the mouse's genome, and this allows for the study of the functions of genes within the entire organism. Genes in mice have been removed, replaced, duplicated, and mutated.

Sometimes, rats are the more appropriate model. Rats share many of the same characteristics as mice, but have the advantage of size, which makes them more suitable for some studies. Although rats share many similarities with mice, rats are not large mice and breeding rats can be its own challenge. This volume will focus on mice, since they comprise by far the largest number of animals bred for research, but rats will be addressed as well. The culture of rat embryonic stem cells and other newly-developed genetic manipulation technologies allowing for manipulation of the rat genome may increase the popularity of rats in research.[3-5]

Combined, the authors have nearly 100 years experience working with both wild and laboratory rats and mice in laboratories, pharmaceutical companies, commercial rodent production, academia, and field studies. We hope that you will find this volume useful in managing your colonies.

PART 1 Origin, history, and behavior

Origin and history

Both rats and mice used in the laboratory today are domesticated animals, as a comparison with wild-caught mice or rats will quickly show. Laboratory mice and rats are fatter, slower, less aggressive, and more amenable to handling than their wild-caught counterparts. As an organism that lives commensally with humans, there have been many opportunities through time for people to establish relationships, good or otherwise, with the small beings living in their homes and fields. Mice originated in the Indian subcontinent and spread throughout the world with agriculture and human movement.[6] The original habitat of the Norway rat is the steppes of northern China and Mongolia, and like mice, rats have spread throughout the world with human migration.[7] *(Figure 1)*

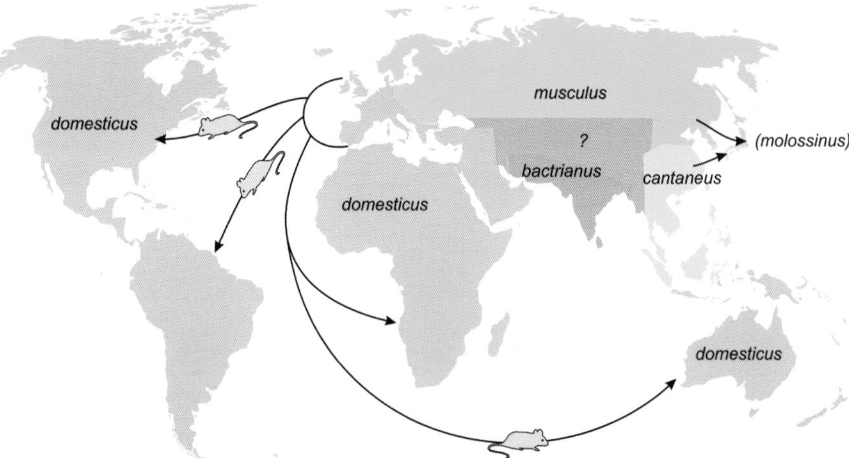

Figure 1: The spread of mice throughout the world from their origin in Asia. Figure after Silver, 1995.[6]

In ancient Asia Minor and Greece, albino mice were sacred to Apollo Smintheus, the god of mice. Mice bred and lived freely in his temple and were used for divination purposes.[8,9] Japanese and Chinese mouse fanciers from 1100 BCE, and perhaps earlier, are responsible for many of the coat color and behavior variations perpetuated by other, later mouse fanciers.[10] These included waltzing mice, spotted mice, chocolate mice, non-agouti mice, and yellow mice. These variations caught the eye of scientists in the late 19th century who wanted to study the new science of genetics and the possible heritability of cancer.

The source of many of the mouse strains currently in use, including the most popular mouse, the C57BL/6, is the mouse colony established by Miss Abbie Lathrop in Granby, Mass.[11] Miss Lathrop was a retired schoolteacher, an enthusiastic mouse fancier, and a scientist in her own right.[12] She imported mice from Europe to breed with her mice, as well as mating them to wild-caught mice from Vermont and Michigan. She supplied mice and collaborated with Dr. C. C. Little, the founder of the Jackson Laboratory, as well as with Dr. William Castle, another pioneer in mouse genetics. Miss Lathrop's colony was not the sole contributor to modern laboratory strains, however. Others included Dr. Clara Lynch, who imported Swiss mice, and the "Bagg albino" from a mouse dealer in Ohio.

Wild Norway rats were captured en masse for the blood sport of rat baiting, popular in England, France, and the United States in the 18th and 19th centuries. In rat baiting, a terrier is put in a pit with wild rats and the number of rats the terrier kills in a certain time is recorded. After being trapped, rats were held in "pounds" in anticipation of the next contest. Albinos were removed from those pounds and held for show or breeding purposes. It is a reasonable assumption in the domestication of laboratory rats that the show rats found their way into laboratories.[13]. It is generally accepted that the rat was the first species domesticated primarily for research purposes.[14] Rats were first used for experiments in the United States in the 1890s, at the University of Chicago, and it is likely those animals were introduced by a newly emigrated faculty member, Dr. Adolf Meyer.[14]

While the history of the laboratory mouse is entwined with the Jackson Laboratory, that of the laboratory rat is linked to the Wistar Institute. The Wistar Institute is where the laboratory rat rose to prominence as a research system, and where many developments were made in housing, husbandry, and health. Wistar origin rats include the WI, BN, LEW, SHR, BDIX, and many others.[14] Wistar Institute scientists gave scientific research the foundation for the use of the rat in many aspects of research, including endocrinology, nutrition, and behavioral research. Dr. William Castle of Harvard and the University of California, Berkeley, a pioneer in mouse genetic research, also played a large role in the foundation of genetic research in rats.

Behavior

When evaluating the normal behavior and biology of laboratory mice and rats, serious consideration must be given to the fact that domesticated animals retain behaviors seen in their wild ancestors. It must also be noted that the descriptions of behavior provided below are meant as general guidelines. Many investigators have noted the myriad strain differences in behavior, and behavior can be modified by environmental factors.

As small, nocturnal prey species, both mice and rats find open spaces aversive, since crossing these open spaces leaves them vulnerable to predators. They are thigmotactic species, meaning that they have a tendency to remain close to vertical surfaces.[15,16] This predilection is exploited by pest control companies, who place bait stations and traps along "runways" used by rodents and at entrances to burrows, both of which are usually located along walls or next to other vertical surfaces. It is also exploited by certain types of behavioral testing, which measure time to cross open spaces, or whether animals will leave dark, enclosed spaces and enter open spaces. In open cages, mice and rats prefer to use the periphery.[17] Both species naturally burrow into the earth, and at least one behavioral test, the visible burrow system, is based on this.[18,19]

Mice and rats are social animals that form stable breeding group units through the formation of dominance hierarchies among both males and females. These stable units are called demes and are usually comprised of related females and males unrelated to these females. Demes occupy territories that both males and females scent mark and defend from interlopers. Gene flow does occur across demes, however.[20] Mice and rats choose mates based on scent and behavior, and mice, given mate choice, have more reproductive success with mates they have chosen.[21,22]

Fighting, especially among males, is a normal part of establishing a dominance hierarchy. Although juvenile mice do not seem to exhibit much play behavior, juvenile rats can be observed mock-fighting and wrestling as part of their normal play. Aggressive interactions, however, usually cease before injury occurs to either animal. When aggression is escalated, wounding and death may occur. Environmental modifications such as shelters and running wheels can affect the intracage hierarchy and their use requires careful evaluation, especially when working with highly aggressive male mice.[23-27] Even nesting material, thought generally to promote affiliative behavior in mice, has been shown to promote aggression in some strains.[28]

Grooming is a normal patterned part of mouse and rat behavior, and large portions of time are allotted to both auto- and allogrooming.[29,30] When grooming behavior goes awry, however, fur barbering, whisker pulling, or ulcerative dermatitis may result. *(Figure 2 a b c)* Fur barbering and whisker pulling may be seen in all strains and stocks of mice and rats, but is most frequently noted in C57BL/6-derived animals, perhaps because they are the most prevalent background used in research.[31,32] Barbering may be an abnormal compulsive behavior initiated by stress (the most likely explanation), part of dominance-related behavior, or part of maternal behavior.[32-34] As vibrissae (whiskers) serve as active tactile sensory organs, animals with barbered whiskers may exhibit abnormal neurobehavioral phenotypes.[35] Barbering behavior may be ameliorated by dietary supplementation or environmental enrichment.[36,37]

Figure 2: A) A normal complement of vibrissae in a mouse.

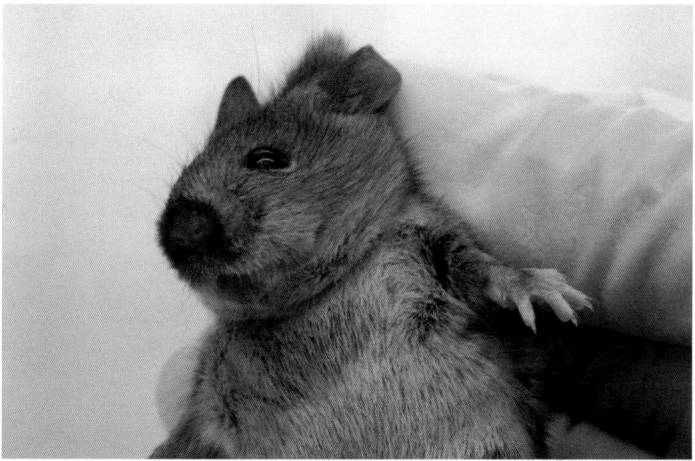

Figure 2: B) An animal whose cagemates have pulled its whiskers out or gnawed them off. This animal may have behavioral and neural changes due to the change in sensory input.

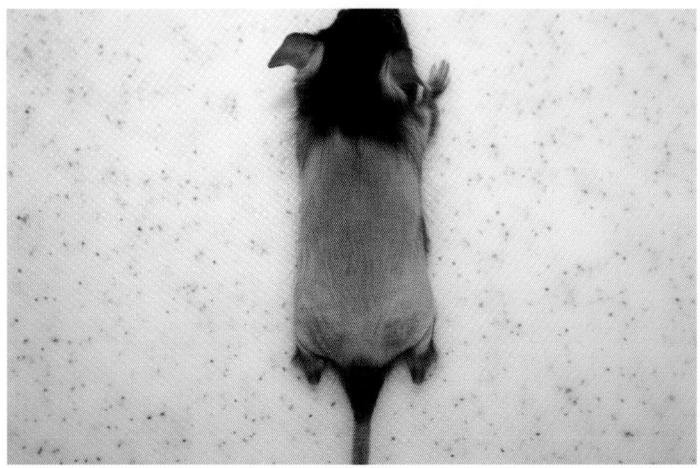

Figure 2: C) A severely barbered mouse. Due to the distribution of the hair loss, this mouse was probably barbered by its mother.

Nest-building is an innate affiliative behavior expressed by both male and female mice. Nests allow mice to live at temperatures that would otherwise prove rapidly fatal, such as in meat lockers. Rats will also build nests, especially dams near term, but nest-building seems to be a learned behavior in rats.[38] Nest-building ability in mice is linked both to ambient temperature and genetic components.[39] Nests can be anything from a shallow cup to an elaborate, domed structure with multiple entrances. *(Figure 3 a b)* In the wild, the outer structure of nests is made of coarse, long-fiber materials, while the inner structure is lined with a softer material. If mice are given more naturalistic nesting materials (something that resembles long fibers, such as hay) they will construct a more naturalistic nest.[37,40] Rats can benefit from the provision of nesting material, especially if they are given an opportunity to learn how to use it.[41]

Stereotypic behavior is repetitive, pointless behavior exhibited by animals, often in response to stress. Stereotypic behavior is seen relatively often in laboratory mice with unenriched environments, but its prevalence varies by strain. Since mice and rats show greater activity levels at night (i.e., dark phase

Figure 3: Nests built by mice. A) shows a complex nest, built with 8 g of long-fiber nesting material (Enviro-dri®, Shepherd Specialty Papers) by a pair of breeding females and a male. The nesting material was provided to the cage 6 days prior to the photo.

Figure 3: B) shows a nest built from aspen shavings by a single mouse in transit for approximately 24 hours.

Table 1. Normal behaviors and their problematic expression in captivity

Normal or wild behaviors	Problematic presentation of these behaviors seen in captivity
Grooming	Barbering, whisker pulling, and ulcerative dermatitis
Aggression	Escalated aggression resulting in severe wounding or death
Nest-building	Huddling in corners
Burrowing	Digging at the cage side
Exploring, foraging and patrolling territory	Bar mouthing and corner jumping, stereotypical movements in the cage such as flipping, circling, or tail carrying
Eating and gnawing	Food grinding

of the light cycle), their behavior should occasionally be observed during the dark cycle. If some animals are showing stereotypic behaviors during the day, when they should be asleep, many more animals may be showing stereotypies at night, when they are normally active. Although environmental enrichment may work to mitigate or prevent some types of aberrant behavior, it may not always be successful.[42] Stereotypic behaviors described in laboratory mice include barbering, food grinding, running on wheels, and patterned movements such as twirling, flipping, corner jumping, bar mouthing, and digging. (*Table 1)* Fewer spontaneous stereotypical behaviors are described in rats; these behaviors in rats are generally induced by administration of drugs. Rats may be less prone to exhibit spontaneous sterotypies, or rat spontaneous sterotypies may be more subtle.

Pheromones, animal genotype, and housing environment can greatly affect animal behavior.[43,44] Stereotypic behavior and other problem behaviors that may occur in barren cages, or when animals are singly housed, can affect not only breeding performance but also research results. The best cure for "bad behavior" is prevention, which is usually accomplished by environmental enrichment. Some suggestions are found in *Table 2*. Mice and rats have different drives and needs for enrichment, and this should be considered when planning an enrichment program.[45-48] In some situations, interaction with humans can be considered environmental enrichment and some rats appear to enjoy such contact.[49] Mice, in contrast, do not seem to seek out social interactions with humans to the same extent as rats, but can be conditioned to tolerate handling.[50] The behavior of both species toward humans benefits from gentle, repeated handling.

Table 2: Enrichment suggestions ranked in order that animals seem to prefer

Mice	**Rats**
Conspecific	Conspecific
Nesting material	Complex space
Complex space	Shelters
Shelters	Gnawing items
Running wheels*	Nesting material**
Human interaction	Human interaction

** Running wheels may promote stereotypic behavior in some strains.*
*** Rats will make greater use of nesting material if exposed early in life.*

PART 2

Biology and reproductive biology

As noted earlier, mice and rats are mammals, and as such share many traits in common with other mammalian species. Below we will discuss where mice and rats diverge from other mammals, and special characteristics that differ between the two species.

Temperature regulation

Both mice and rats, but especially mice, have a relatively large surface area per gram of body weight. This makes them more sensitive to cold than larger animals. For mice, their typical response to cold is non-shivering thermogenesis or behavioral modification, such as burrowing or nest-building. When not given the means to adapt to cold conditions, mice will show stress at temperatures below 18°C. Note that mice and rats are typically housed in ambient temperatures between 20 -24°C, which are comfortable for people working in animal rooms. A wide estimation of the mouse's thermoneutral zone is 26-34°C based on published literature addressing many mouse strains and with a wide variety of methodologies.[51] The lower end of this thermoneutral zone for any particular mouse is probably closer to 30°C[52], which means that mice are chronically cold stressed in typical laboratory housing. One relatively recent study suggests a similar situation for rats, with their chosen housing temperature being 25-27°C when housed on a thermocline of 15-40°C.[53] Neither rats nor mice pant, nor do they sweat. Their only recourse when temperatures become too hot are again, behavioral adaptations, such as burrowing, or wetting their fur with saliva for evaporative cooling.

Urinary system

Although they can conserve water through urine concentration, mice and rats produce a large amount of dilute urine when compared to true desert dwellers such as gerbils. Mice and rats normally excrete large amounts of protein in their urine. These are mainly lipocalin proteins that transport pheromones and are the primary allergens produced by mice and rats (Mus m 1 in mice and Rat n 1 in rats).[54] Mouse and rat urine also fluoresces in the UV range, which may be used by the animals for navigation and by predators to spot rodent trails.

Digestive system

Both rats and mice are monophydont hypsodonts, meaning that they have one set of open-rooted teeth. Their dental formula is identical: (1/1 0/0 0/0 3/3), with two upper incisors, two lower incisors, and three pairs of upper and lower molars. The enamel of rodent incisors incorporates iron, making it very durable and giving it an orange-yellow cast. Mouse and rat teeth grow continuously, due to their open roots, and although much of the tooth wear occurs at the occlusal surfaces, gnawing by animals also serves to wear teeth. Enamel is present on the labial surfaces of the teeth, while the lingual surfaces are primarily dentin. Mouse incisors grow at a rate of approximately 2 mm/week for upper incisors and 2.8 mm/week for lower incisors.[55] Rat tooth growth rates are quite similar: 2.2 mm/week for maxillary and 3.0 mm/week for mandibular incisors.[56] The incisors erupt at postnatal day 9-12 in both species.

Although rodents in general are often classified as a group as herbivorous animals, rats and mice are omnivores. For example, wild mice and rats will supplement a diet of grains and seeds with human rubbish, fruit, insects, amphibians, birds, eggs, carrion, and other mice or rats. This may mean that they have nutritional requirements that are not being adequately met by commercial rodent diets. Cannibalism does occur and is a normal response to dead pups in the nest or normal investigative behavior when confronted with a dead cagemate. Both mice and rats have the large cecum typically present in hindgut fermenters that allows them to digest cellulose. Mice have gallbladders; rats do not. Both species are coprophagic, and this behavior is seen especially when young animals begin to eat solid food. This may help the young establish necessary gut flora.[57]

Muskuloskeletal system

A typical vertebral formula for a mouse or rat is: C7 T13 L6 S4 C28. Strain variations occur, especially in the thoracic, lumbar, and caudal regions.

Special senses

To better understand the behavior of mice and rats, it helps to understand how they experience the world. Their senses differ from those of humans in sensitivity and primary sensory modality. Although far from the human experience, understanding how mice and rats process their interactions with us through their senses provides valuable input as to how we can adapt environments to suit them. *(Figure 4 a b)*

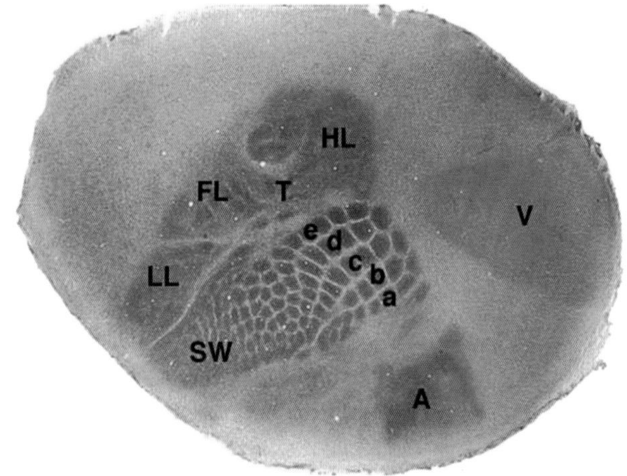

Figure 4: A) An illustration of the mouse's sensory cortex. This figure shows the flattened left hemisphere of a GAP-43 wildtype mouse cortex immunostained with serotonin transporter. The layer IV thalamocortical afferents show a complete body map: V: visual, A: auditory, a-e: large whiskers, arranged in 5 rows on the snout, SW: small whiskers, LL: lower lip, FL: forelimb, HL: hindlimb, T: trunk. Note the huge component of the sensory cortex devoted to the whiskers. Photo used courtesy of Dr. S. Donovan and Dr. J. McCasland (*Proc. Natl. Acad. Sci*. 1999: 96, 9397-9402)

Figure 4: B) If the sensory input into a mouse's cortex is drawn as a figure, with the size of the body part represented by the size of the cortex devoted to that body part, this is how the mouse would appear. The dominant sensory modalities for a mouse are very different than those of a human. (Drawing after an illustration by Steve Moskowitz, Advanced Medical Graphics.)

Vision: As nocturnal prey species, rats and mice have vision adapted to see in dim light and see motion more than form, meaning they have poor visual acuity but a great depth of field and are very sensitive to any motion in their visual field. Rods, the retinal cells that distinguish light from dark, predominate in both species, comprising ~99% of the cells in the retina, compared with ~95% in humans. Both species are dichromats, meaning they have two types of cones (color vision cells) in their retinas. In comparison, humans have three types of cones. This gives mice and rats different color vision than humans and vision sensitivities at different parts of the spectrum. For example, they see wavelengths we cannot in the ultraviolet but are significantly less sensitive to red light, which is greater than 600 nm. Their vision is most sensitive at ~505 nm (between blue and green). *(Figure 5)* When compared to humans, rats and mice have what humans would consider very poor eyesight.[58-60] Thus, many of the cues used in behavioral tests may be neither visible nor interpretable by animals. Albino rats and most sighted mice have a visual acuity of 0.5 cycle/degrees (c/d) and pigmented rats ~1 cycle/degree. To convert that to typical measures

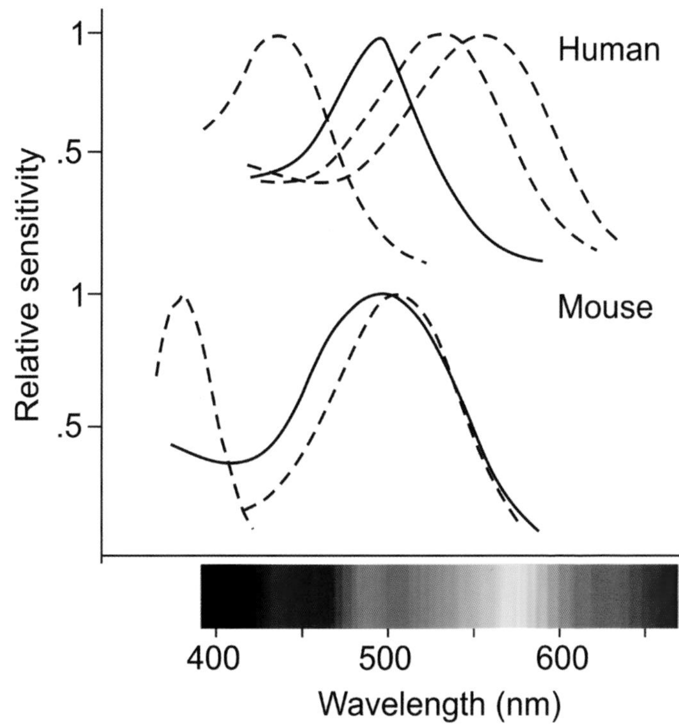

Figure 5: An approximate representation of how human sensitivity to light differs from that of mice. The solid lines illustrate the light sensitivity of rod cells and the dashed lines represent cone cells. Note that mice only have two types of cones and that their rods are sensitive over a longer portion of the spectrum than humans. This also illustrates the ability of mice to see into the ultraviolet.

of human vision, they are extremely nearsighted. Human 20/20 vision under standard testing conditions converts to 30 c/d. Using that conversion factor, mice and albino rats have 20/1200 vision and pigmented rats have 20/600 vision.

For mice to have poor vision, they must have sight at all. Many retinal degeneration alleles are fixed in laboratory mouse populations.[61] The most widespread is the allele formerly known as retinal degeneration (*rd* or *rd1*; now called *Pde6b*^{rd1}). This gene causes an early onset retinal degeneration, meaning animals lose their rod cells by 35 days of age. Blind strains include C3H/HeJ, CBA, FVB, SJL, P, and PL and blind stocks include Black Swiss, some ICR, Swiss Webster, and NIH.[62,63] A review of the literature does not reveal any retinal degeneration alleles fixed in a wide variety of laboratory rat populations, although RCS rats are known to carry a mutation for retinal dystrophy.[64] Albino animals may also be susceptible to blindness from light levels in animal rooms. Their lack of pigmentation, as well as reduced numbers of rod receptors associated with the albino mutation, leaves them uniquely vulnerable to inadvertent light damage.[65,66]

Hearing: Many rodents have peak auditory sensitivities in the range of 15-30 kHz compared to the human ear's peak sensitivity at 2-4 kHz. Ultrasonic frequencies, as defined by human hearing, begin at greater than 20 kHz. A mouse's hearing range is from ~1 kHz to 90 kHz, with peak sensitivity at 10-22 kHz, while rats can hear from 0.5 to 64 kHz, with peak sensitivity between 6-42 kHz.[67] *(Figure 6)* Mice and rats communicate in ultrasonic frequencies. Pups call to dams[68], rats laugh[69], and male mice sing mating songs to females.[70,71] This ultrasonic sensitivity means that mice and rats can be

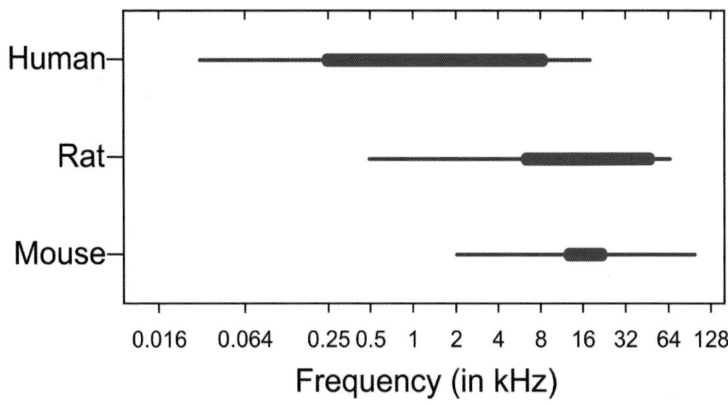

Figure 6: A comparison of mouse, rat, and human hearing ranges. The thin lines indicate total hearing range (usually considered to be frequencies that can be heard at 60dB) and the thick lines show the peak sensitivities, which are those frequencies audible at 10dB. Figure after Heffner and Heffner, 2007.[67]

disturbed by noises beyond human hearing, and conversely, noises we find irritating, they may not hear at all. Potential noise disturbances in the ultrasonic include cage washers, oscilloscopes, computers, video display terminals, running water taps, and nearby construction. Some mouse strains, including A/J, DBA/2J, and NOD/ShiLtJ, carry a recessive mutation that causes age-related hearing loss that may occur as early as 2 months of age.[72] Since high-frequency hearing is lost first, these strains may be at a disadvantage when it comes to communication with other mice.

Smell: Rodents have a keen sense of smell. Their olfactory cortex is a large proportion of their sensory cortex and is contained in two bulbs at the rostral end of the brain. Olfactory cues are important components of social, sexual, and parental behaviors. Urinary, fecal, salivary, preputial gland, and plantar gland odors communicate relatedness, health, and dominance, as well as recent fright. Pheromones are detected by the vomeronasal organ, part of the accessory olfactory system. This system does not feed through the olfactory bulbs to the olfactory cortex, but rather connects via the accessory olfactory bulbs to the amygdala, the stria nucleus, and the hypothalamus. For laboratory mice, scent marking of cages is probably an important part of the behavioral repertoire after cage change although laboratory rats do not seem to be disturbed by cage changing.[73] In mice, the transfer of nesting material with its affiliative pheromones at cage change has shown to decrease fighting when animals are placed in clean cages.[74]

Touch: Besides the usual mechanoreceptors present in haired skin,[75] mice and rats have specially adapted hairs, called whiskers or vibrissae. Located primarily on the rostrum, vibrissae may also be found on the head, in the cervical region, and on the feet. The part of the sensory cortex associated with inputs from whiskers is enormous in comparison to other sensory afferents. Interacting with the environment via their whiskers is integral to mouse and rat behavior. Whisking is an active activity[76] and animals can discriminate between subtle tactile differences with their vibrissae.[77,78] Rodent vibrissae are probably best considered as equivalent to human fingertips in terms of their sensitivity and cortical input.

Reproductive biology

Mice and rats are litter-bearing mammals. They are spontaneous ovulators, have short gestation periods, and may reproduce year round. The overall reproductive performance of mice and rats varies widely, with some stocks and strains completely infertile due to problems with one or both sexes, and others fertile for their entire lifespan. Most stocks and strains are somewhere in between, with some background level of infertility in both males and females, and most animals reaching reproductive senescence before the end of their lifespan. It is important to remember that reproductive performance is highly dependent on genetic background, with outbred animals generally being more fecund than inbred. Detailed information on the reproductive biology of mice and rats may be found in Pritchett and Taft,[79] Lohmiller and Swing,[80] Hardy,[81], Maeda et al,[82], and Zimmerman et al.[83]

Basic reproductive hormonal function follows a typical mammalian pattern with multiple hormones produced by the brain and the reproductive tissues, all of which interact in complex ways. *(Figures 7 and 8)* The hypothalamus releases gonadotropin releasing hormone (GnRH) that acts on the anterior pituitary. The anterior pituitary then releases follicle stimulating hormone (FSH) and lutenizing hormone (LH). These two hormones act on the gonadal tissues that in turn produce testosterone, estrogen, and progesterone. These feed back to the hypothalamus. In the female, FSH released by the pituitary stimulates the development of ova in the ovaries. The ovary produces estrogen in response to FSH, and this results in estrous changes in the animal. Estrogen is responsible for the LH surge that then results in the maturation and release of ova. After ova are released, corpora lutea (CL) form. CL produce progesterone, which maintains pregnancy. Mice and rats are spontaneous ovulators but a copulatory stimulus is necessary to support the CL regardless of fertilization status. Without this copulatory stimulus, the CL

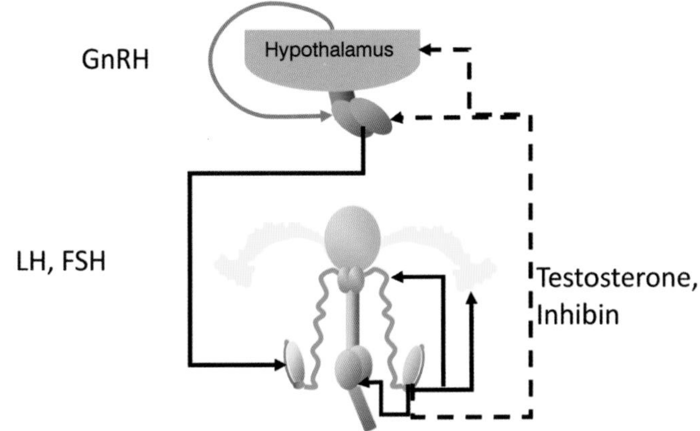

Figure 7: Hormonal control of male reproductive function. Lines indicate hormonal loops between the structures, with solid lines indicating positive feedback loops and dashed lines indicating negative feedback loops. Figure after Dr. R. Taft.

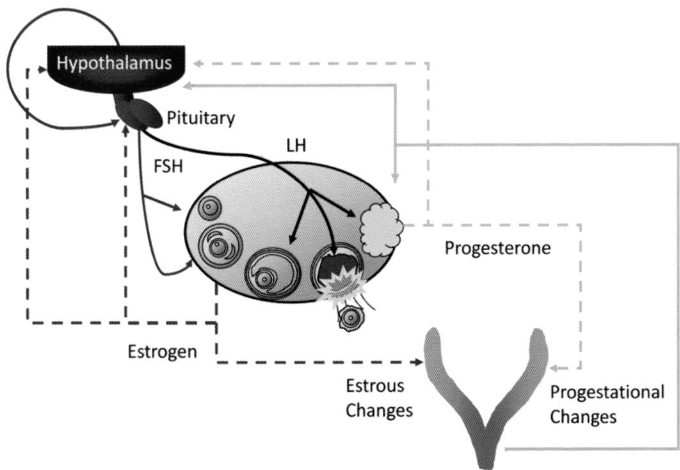

Figure 8: Hormonal control of female reproductive function. Lines indicate hormonal loops between the structures, with solid lines indicating positive feedback loops and dashed lines indicating negative feedback loops. Figure after Dr. R. Taft.

will degenerate. With it, the animal will sustain a pregnancy or remain pseudopregnant for 10-13 days.

In males, GnRH is released by the hypothalamus and the pituitary also releases FSH and LH. Both hormones are required for the initiation of spermatogenesis, but not necessarily for the continuation of spermatogenesis once it is initiated. FSH and LH act to establish a full complement of cells (Sertoli, Leydig, and spermatogonia) that are necessary to the proper functioning of a mature testis. Both testosterone and FSH are necessary for further development of spermatogonia into spermatids.[84]

Chromosomal sex is determined in the embryo at conception, but gonadal sex is apparent at midgestation in the mouse and rat. Gonadal sex is determined by the presence or absence of certain genes, and the default phenotypic presentation is female, regardless of chromosomal composition. Animals can be sexed at birth, if a particular litter composition is desired or animals of one sex are experimental subjects. *(Figure 9)* Before the descent of the testes, sex is most easily determined based on the distance between the genital papilla and the anal opening (anogenital distance). As the hair coat erupts at approximately day 5, it becomes easier to see the nipples in females. Anogenital distance in males is about 2 times that of females. Factors that can affect anogenital distance include position of the pups *in utero* and exposure to endocrine-disrupting chemicals.[85,86] Males with an in utero position between two females have a more feminine anogenital distance than those between two males while females between two males *in utero* are masculinized.[87]

Male anatomy

Mice and rats have practically identical male anatomy. *(Figure 10)* Male animals have paired testes and their associated structures, the epididymides and vasa deferentia, a single penis, paired seminal vesicles (or vesicular glands) with coagulating glands tightly adherent to them, a dorsoventral prostate, bulbourethral glands, and preputial glands, located subcutaneously in the ventral abdomen. Male mice and

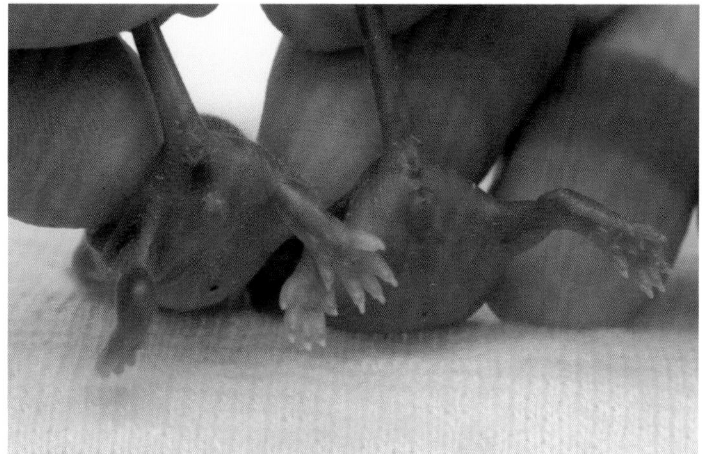

Figure 9: Photographs of mice and rats at various ages, illustrating the differences between the sexes. In every photograph, males are on the left.
A) Newborn mice

Figure 9: B) Newborn rats

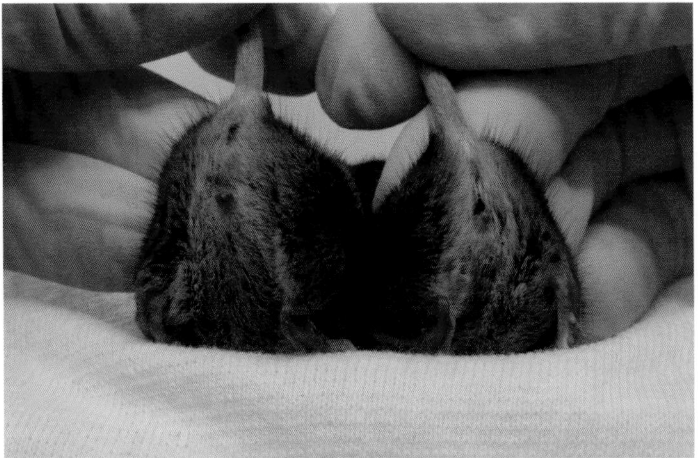

Figure 9: C) Weanling mice

Figure 9: D) Weanling rats

Figure 9: E) Adult mice

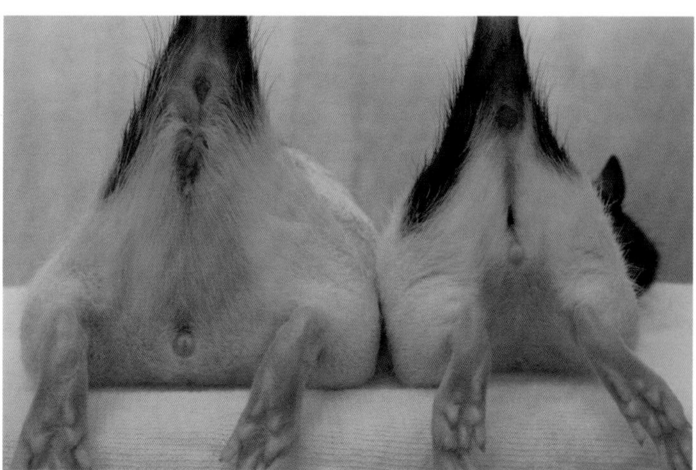

Figure 9: F) Adult rats.

rats have open inguinal rings, so if castration is necessary, the rings must be closed. Neither male mice nor rats have nipples, due to androgen-mediated destruction of the mammary bud at E14.[88]

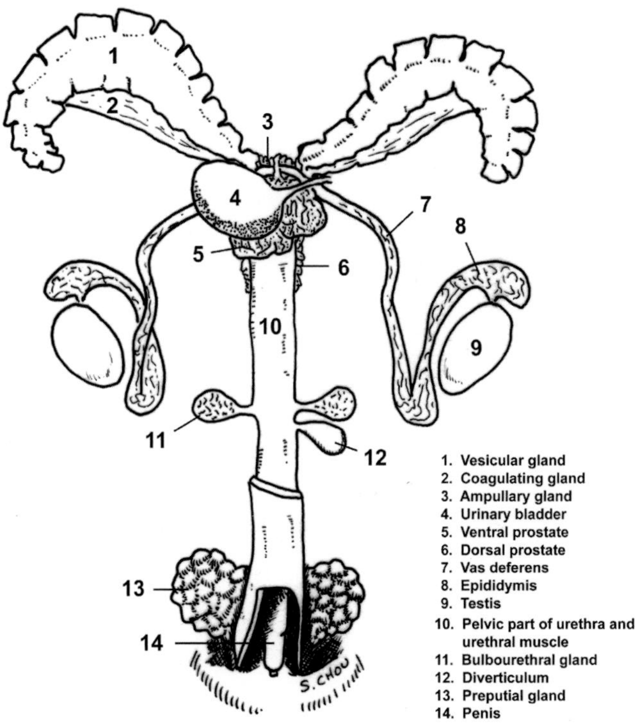

Figure 10: Male mouse reproductive anatomy. Figure from Pritchett and Taft, 2007[79], used with permission.

A normal mouse or rat has continuous germ cell differentiation in the seminiferous tubules of the testes. There are four stages of sperm development with a maximum interval of approximately 35 days for the differentiation process. Spermatogonia (diploid germ cells) become spermatocytes (diploid cells) that in turn become spermatids (haploid cells) and further differentiate into spermatozoa (morphologically mature haploid germ cells, "sperm").[89] From the testis, spermatozoa are released into the epididymis to undergo further biochemical changes and become fully functional sperm. Sperm are stored in the epidiymis until ejaculation. *(Figure 11 a b).*

Sperm production and quality varies by strain.[90] Libido recovery (willingness to breed again after successful coitus) also varies by strain.[91] Recovery of an animal's libido can occur in as short as 2-4 hours or may take as long as two weeks.[92] Sperm production does not correlate with libido recovery, so it is possible to have animals that are willing to mate, but have very few sperm. A good guideline is to allow at least four days between mating events for outbred males used intensively and seven days for inbred males. The copulation plug is a yellow to white plug produced by secretions of the coagulating gland, seminal vesicles, and prostate. It typically fills the vagina and provides a mechanical barrier, although not an absolute one, to mating by another male.[93] The presence of a copulation plug indicates successful mating and ejaculation, but does not necessarily indicate pregnancy.[94] Copulation plugs are retained longer in mice than in rats and it is wise to check the bedding of reats for the presence of a plug. *(Figure 12 a b c)*

Female anatomy

The overall anatomy of the female mouse and rat is very similar. *(Figure 13)* The female reproductive tract is comprised of ovaries and their related structures, the oviducts, the uterus, the vagina, and the vulva. The paired ovaries are surrounded by ovarian bursae and attached to the uterus by the oviducts. The uterus is bicornuate with a single cervical os in the vagina. The vulva is composed of the vaginal introitus,

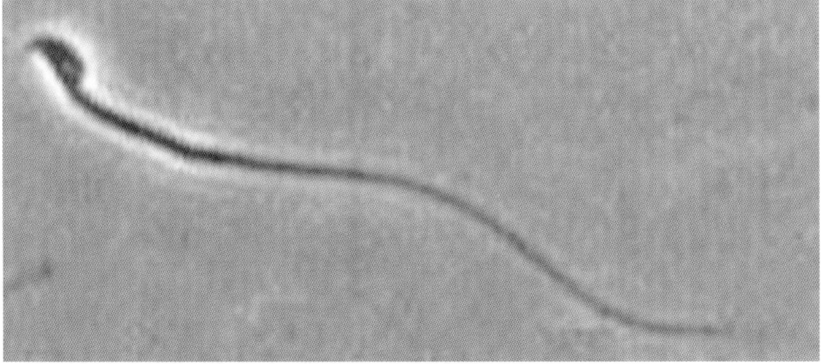

Figure 11a: A normal spermatozoon from a mouse

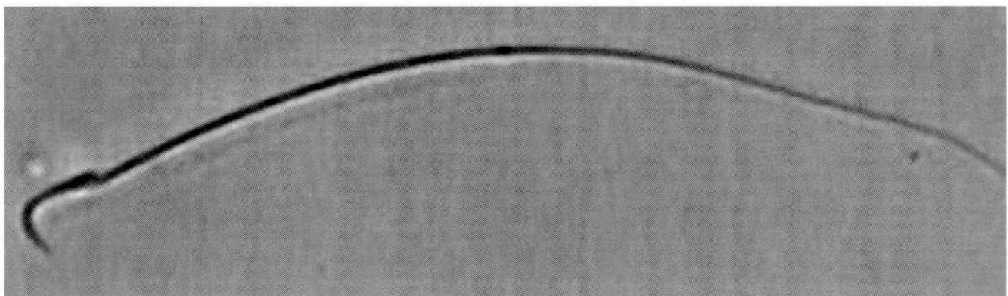

Figure 11b: A normal spermatozoon from a rat

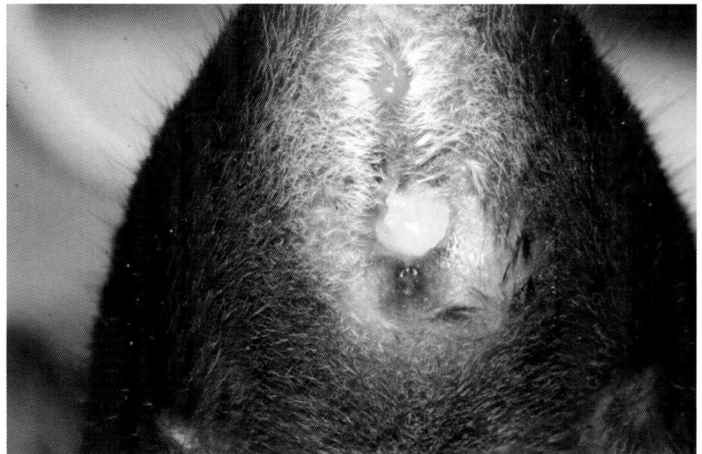

Figure 12A: Copulation plug in a mouse

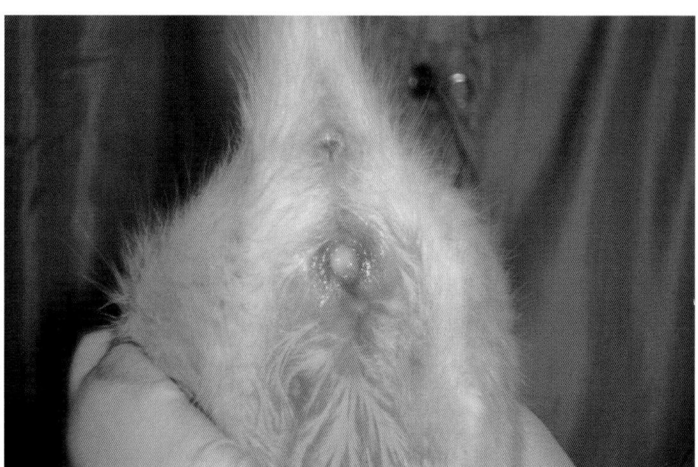

Figure 12b: Copulation plug in a rat

Figure 12C: Copulation plugs on the floor of a rat cage. Copulation plugs do not tend to remain fixed in rats for as long as they do in mice, so if examining rats for copulation plugs, check the bedding as well.

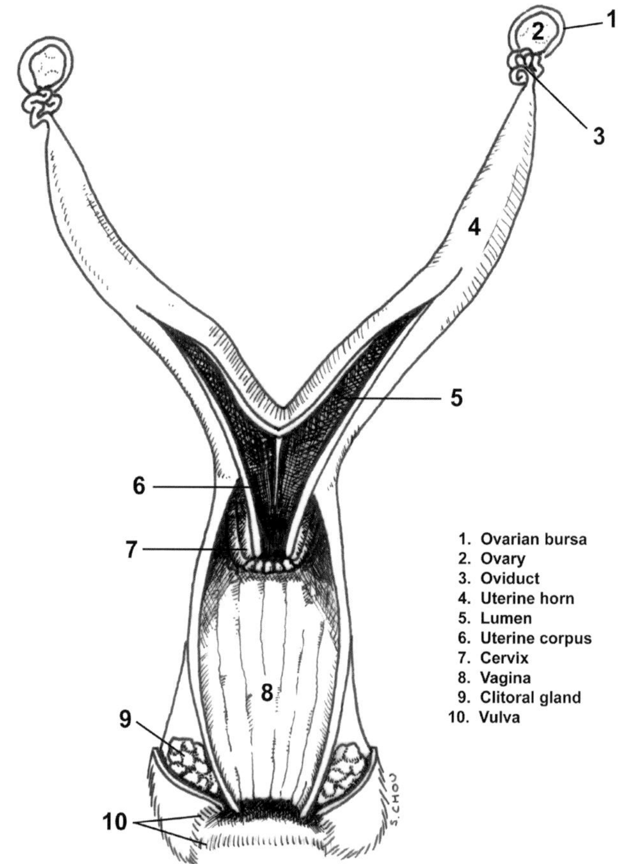

Figure 13: Female mouse reproductive anatomy. Figure from Pritchett and Taft, 2007[79], used with permission.

the clitoral glands, and the urinary papilla. Mammary tissue is present in two distinct regions—cervicothoracic, where it can extend around to the dorsum, and inguinoabdominal. Mice have 5 pairs of mammary glands and teats; 3 pairs in the cervicothoracic region and 2 pairs in the inguinoabdominal region. Rats have 6 pairs of mammary glands and teats; 3 pairs in the cervicothoracic region and 3 pairs in the inguinoabdominal region. *(Figure 14 a b)*

A female mouse or rat is born with all of the oocytes that she will ever have, a total of approximately 30,000–75,000.

Oocytes remain quiescent until sexual maturation. During each natural cycle, only 6-16 oocytes are hormonally stimulated to undergo ovulation. After mating takes place (during late proestrus and early estrus), the fertilization of eggs takes place in the oviduct. The egg remains viable for 10-15 hours after ovulation. The vast majority of the oocytes

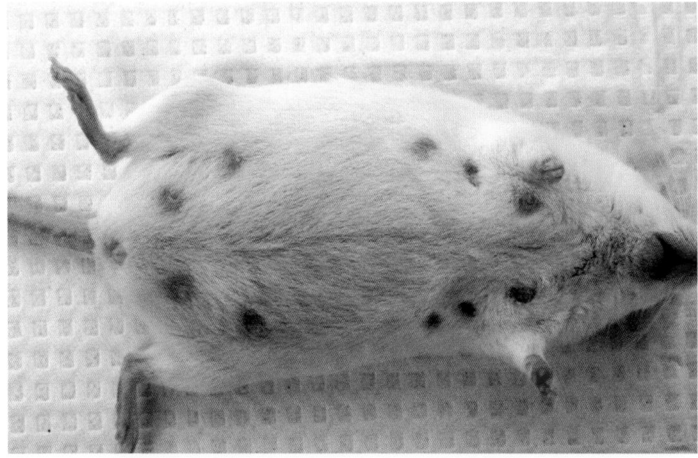

Figure 14a: A nursing mouse, illustrating typical mammary development and the number of mammae.

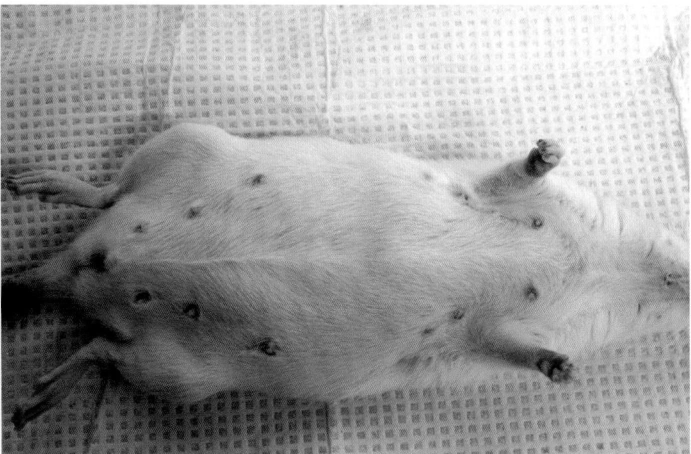

Figure 14b: A nursing rat, illustrating typical mammary development and the number of mammae.

are never ovulated and degenerate throughout the life of the animal.

Mice and rats are polyestrous spontaneous ovulators that cycle every 4-5 days.[95] The estrous cycle follows a typical pattern, beginning with diestrus, followed by proestrus, then estrus, then metestrus. Estrous stage can be assessed by visible changes in the vulva and introitus as well as by smears of vaginal epithelium.[96,97,98] *(Tables 3 and 4 and Figures 15, 16)* Both mice and rats have a fertile postpartum estrus that occurs approximately 12 hours after giving birth. The exact time of estrus is dependent on both circadian rhythms and

Table 3: Classification of the stages of the estrous cycle by cell morphology in vaginal smears.
Table adapted from Nelson, 1982[97]

Stage of cycle	Cell type Leukocytes
Diestrus/proestrus	+ to ++ (Predominant)
Proestrus	0 to +
Proestrus/estrus	0
Estrus	0
Metestrus 1	0 to ++
Metestrus 2	++ to +++ (Predominant)
Diestrus	+ to +++ (Predominant)

Cell density:
0 = none, + = few, ++ = moderate, +++ = heavy

time since parturition.[99-101] Delayed implantation of fertilized ova is possible; this phenomenon is known as embryonic diapause.[102] If a pregnant female is stressed, such as by heavy lactation, embryos can be held as hatched blastocysts in the uterine lumen for as long as 12 days.[103]

Control of the estrous cycle in mice and rats can be achieved in many ways. Exogenous hormones are the simplest way and this method is used extensively in transgenic cores. Response to exogenous hormones depends on genetic background as well as timing of administration.[104,105] Both rats and mice will synchronize estrous cycles with other females.[106,107] Individual

Cell type		
Nucleated epithelia	**Cornified epithelia**	**Smear density**
+ Well-formed	0 to +	Thin
+ to +++ Well-formed (Predominant)	0 to +	Medium
+ to ++	++ to +++ (Predominant)	Medium
0	++ to +++ Relatively small cells (Predominant)	Medium to heavy
0	++ to +++ Larger, more flat and clumped than in estrus (Predominant)	Medium to heavy
+ to ++ Often irregularly shaped and vacuolated	+ to ++	Medium to heavy
+ Often irregularly shaped and vacuolated	0	Thin

Figure 15: Cytologic illustration of the estrous cycle in the mouse. All photomicrographs are taken at 400x and all are stained with Diff-Quik® (a modified Wright's stain).

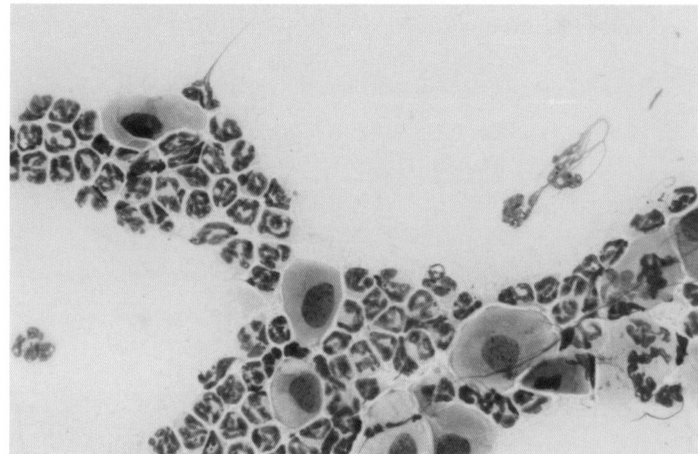

A) diestrus, characterized by nucleated epithelial cells and large numbers of neutrophils

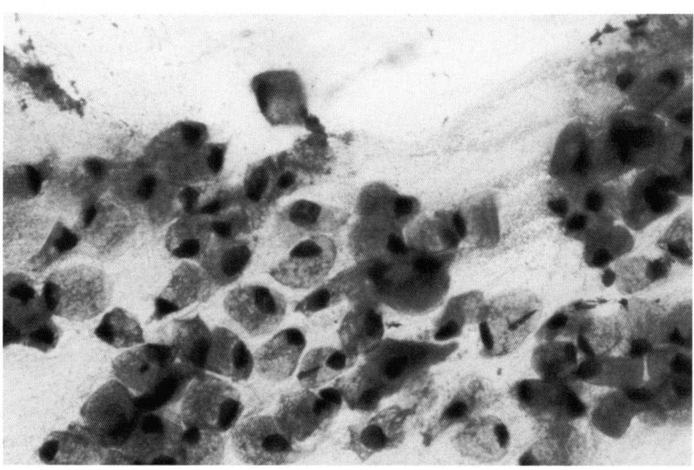

B) proestrus, characterized by nucleated epithelial cells as the primary cell

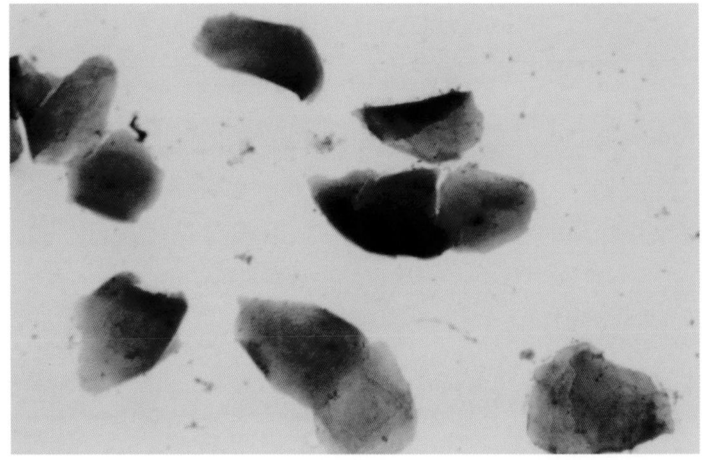

C) estrus, characterized by cornified epithelial cells as the dominant cell type and

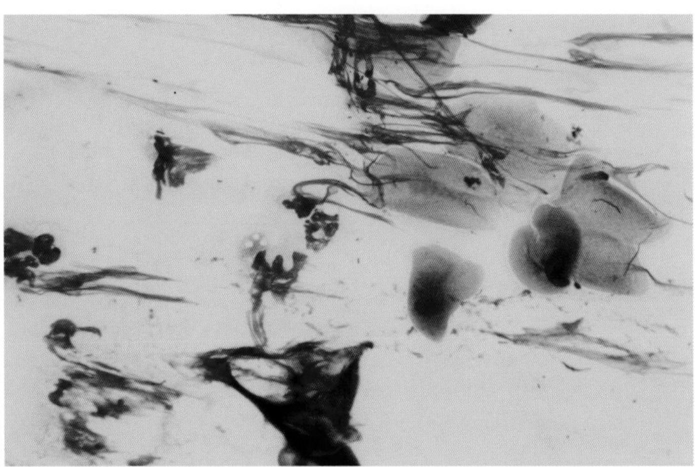

D) metestrus, characterized by the presence of all three cell types, nucleated epithelial cells, neutrophils, and cornified epithelial cells.

Table 4: Appearance of the vagina at various stages of the estrous cycle. Table adapted from Champlin, Dorr, and Gates, 1973.[96]

Estrus stage	Appearance
Diestrus	Vagina has a small opening and the tissues are bluish-purple in color and very moist
Proestrus	Vagina is gaping and the tissues are reddish-pink and moist. Numerous longitudinal folds or striations are visible on both the dorsal and ventral lips
Estrus	Vaginal signs are similar to proestrus, but the tissues are lighter pink and less moist, and the striations are more pronounced
Metestrus 1	Vaginal tissues are pale and dry. Dorsal lip is not as edematous as in estrus
Metestrus 2	Vaginal signs are similar to metestrus-1, but the lip is less edematous and has receded. Whitish cellular debris may line the inner walls or partially fill the vagina

housing of mice will suppress the estrous cycle, as female mice need exposure to pheromones present in male urine to cycle regularly. Overcrowding will also suppress estrus in mice and this is known as the Lee-Boot effect.[108] The estrous cycle lengthens and the animals remain in diestrus. For the The Lee-Boot effect to completely supress the estrous cycle in all animals, the cage must be overcrowded. The Whitten effect is the induction of estrus in female mice by exposure to the urine of male mice.[109] The Whitten effect works best when combined with the Lee-Boot effect. Overcrowding mice to suppress the estrous cycle, then separating them and exposing them to male urine is a low-tech way to synchronize estrus in a group of mice. Finally, the Bruce effect is the failure of implantation in female mice exposed to unfamiliar males.[110] The effect only occurs if exposure to the new male is before implantation (before E5).[111] Since the effect is mediated via pheromonal components of the urine,[112] it works best when the strange male is of a different strain.[113] Many of these effects do not work in rats, however. The Bruce and Whitten effects are absent and the Lee-Boot effect is milder.

A summary of the basic reproductive parameters of mice and rats can be found as *Tables 5 and 6.*

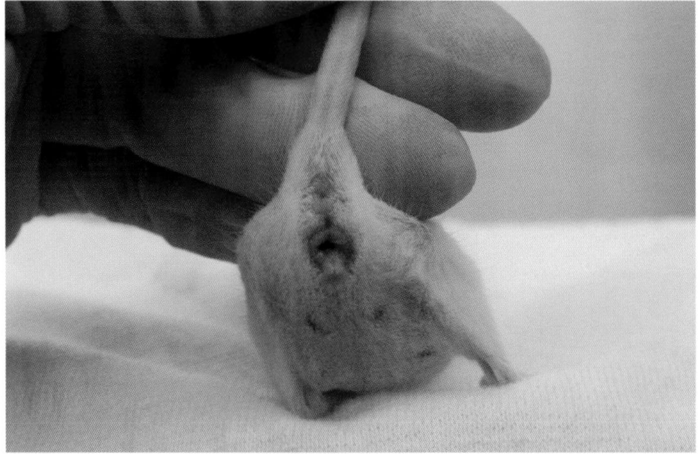

Figure 16: A) The appearance of the vulva of a mouse in proestrus. The introitus is gaping and the tissues are reddish-pink and appear swollen. Longitudinal folds are visible on both the dorsal and ventral lips.

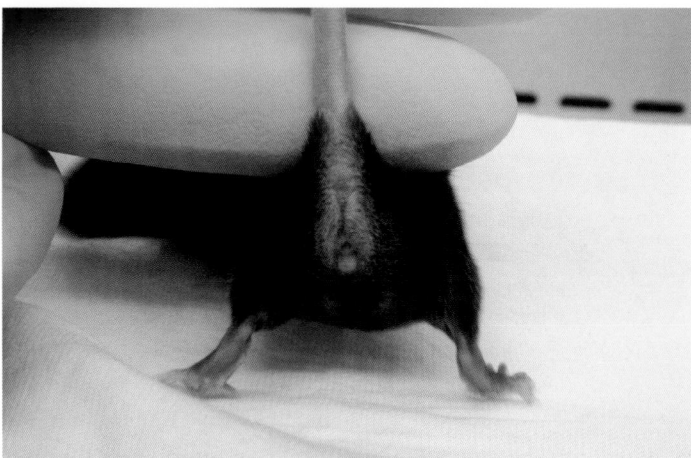

Figure 16: B) A mouse in diestrus has a small introitus and the tissues are blue or purple. There is little to no swelling.

Table 5: Summary of mouse reproductive data

Weight at birth	1 g
Weight at weaning	10-15 g
Weight of adult	20-40 g
Age at weaning	18-28 days
Puberty	5 weeks
Full sexual maturity	7-9 weeks
Estrous cycle length	4-5 days
Sexual receptivity during estrus	12 hours
Fertilization	2 hours after mating
Formation of blastocele	2-4 days
Implantation	4-5 days
Pseudopregnancy duration	10-13 days
Gestation period	19-21 days
Fertile postpartum estrus	Yes
Litter size	6-12
Incisors erupt	9-10 days
First solid food intake	11-12 days

Table 6: Summary of rat reproductive data:

Weight at birth	5-6 g
Weight at weaning	40-50 g
Weight of adult	150-500 g
Age at weaning	21-28 days
Vaginal opening	33-42 days (approximately 100g)
Puberty	~7 weeks
Full sexual maturity	8-10 weeks
Estrous cycle length	4-5 days
Sexual receptivity during estrus	12 hours
Fertilization	2 hours after mating
Formation of blastocele	2-4 days
Implantation	4-5 days
Pseudopregnancy duration	12-14 days
Gestation period	20-24 days
Fertile postpartum estrus	Yes
Litter size	4-15
Incisors erupt	9-10 days
First solid food intake	11-12 days

Mating

The courtship and mating behavior of mice and rats generally follows a consistent pattern. There is a distinct sequence of approach by the male, acceptance by the female, mating, ejaculation, and a refractory period.[92,114,115] Males generally investigate females by sniffing the genitals. If females are not receptive, they avoid contact with males, and if mounting is attempted, they may assume defensive postures. In female rats, receptivity to the male is indicated by ultrasonic vocalizations, hopping, ear wiggling, and lordosis.[116] This is followed by the male mounting, intromission, ejaculation, dismounting, grooming of the genitals, and a refractory period. This sequence may repeat multiple times during proestrus and estrus. Generally, the female is not receptive to the male except during estrus.

Length of gestation

Gestation length is controlled by both genetic and environmental factors. For example, due to the phenomenon of embryonic diapause, the gestation period may appear to be as long as 34-38 days. However, it is generally 18-21 days for mice and 20-23 days for rats. In inbred mice, gestation length has been found to be highly strain-dependent.[117] There is at least some component of the length of gestation that is litter-size-dependent, with large litters born earlier.[118] Fetuses may be palpable by 10-12 days post-conception in mice and 9-10 days in rats—gently feel for growing "beads" in the lower abdomen. Firm palpation can result in lysis of the embryo, terminating the pregnancy.

Parturition

Parturition occurs when the fetuses indicate their readiness. Corticosteroid hormones secreted by the placenta induce luteolysis and begin the parturition sequence. The female exhibits nesting behavior before parturition, and the nests built at that time are complex and differ from nests built at other times.[119] Parturition usually occurs at night for mice, but during the day for rats.[120,121] Parturition takes place over a variable span of time, with a pup being delivered every few minutes and with the entire litter delivered over the course of 1-3.5

hours.[115,122] If undisturbed, the female will lick and clean the pups between each birth and remain in the nest. If disturbed, she may exit the nest between births.

Parental behavior and rearing pups

Infanticide is a normal reproductive strategy in mice and rats.[123-127] Both males and females, and whether virgin or experienced, will commit infanticide. It is commonly seen as a male reproductive strategy however, since killing a litter removes the lactational block to estrus and also removes a rival's genes. Females will also kill pups if stressed or if resources are limited.[128] In mice and rats, the timing of copulation and remaining in contact with a female results in a change in male behavior from infanticidal to parental at about the time offspring from that mating would be born.[124,127] Effective nursing of pups requires synthesis of milk in the mammary glands, parental retrieval of the pups to the nest, crouching behavior by the mother, attachment of the pups to the teats, suckling behavior by the pups, and milk ejection.[119,129] Lactation is a physiological stressor for animals which can suppress estrus or prevent blastocysts from implanting. Nursing positions in mice and rats are related to stress levels. Animals that are fearful or disturbed will nurse in a position that covers all their pups with their body, while animals that are not stressed will nurse in a variety of positions including standing, side-lying, and semi-sitting. *(Figure 17)* Retrieval behavior requires participation and feedback from the pups. Infant rodents emit high-frequency vocalizations that guide the mother to their location and facilitate retrieval to the nest.[130]

For successful growth of pups, adequate milk production must occur. A certain number of suckling offspring is necessary for adequate milk production for all offspring. This number appears to be related to suckling stimulus provided to the mother during the first day of lactation. Peak lactation in rats and mice occurs between days 10-16.[82,131,132] Female mice and rats will nest communally and also communally nurse their pups.[133] This behavior is more commonly seen in mice because female rats are generally not

Figure 17: An example of a relaxed nursing posture in a rat. One pup is nursing and the rest of the litter is resting in the foreground.

housed in pairs or groups. Female mice will make a common nest where all pups are housed. The females enter the nest, nurse the pups that signal a desire to nurse, and then exit the nest. Females will preferentially nest with and nurse the offspring of related mice, but familiarity with another mouse can substitute for relatedness.[134]

For information on the postnatal development of mice and rats and how development can be used to age pups, see *Tables 7 and 8*. Mice and rats are generally weaned at approximately 21 days of age. Some genetically modified animals and inbred strains are smaller and animals may benefit from a later weaning date, especially if the facility has automatic watering. However, this can lead to overcrowding in the cage, since a subsequent litter is likely to follow along ~19-22 days after the first. Some facilities recommend weaning young mice into cages with "aunts" (retired female breeders). These females must be removed from the cage before young males reach sexual maturity. Some hybrid and outbred mice are quite large and reach sexual maturity closer to 4 weeks than 6. It may be best to wean these animals at 2.5 weeks or to remove the adult male from the cage to avoid inadvertent parent/offspring matings.

Reproductive lifespan and breeding unit replacement criteria

Age at sexual maturity can vary depending on strain and environmental conditions. In general, female mice and rats attain puberty, as defined by vaginal opening, between 4-6 weeks of age, and sexual maturity at 6-8 weeks of age. Male mice and rats reach puberty, as defined by sperm found in the tail of the epididymis, at approximately 5-7 weeks.[135,136] Many investigators give animals an extra week or two to mature to attempt to maximize pup survival. This would mean breeding pairs are set up at 7 weeks for mice and 9 weeks for rats.

In general, it is best to retire animals 6-8 months after they are placed together to breed. Male reproductive lifespan is longer, especially if they are housed with females,[137,138] but in commercial production settings and when animals are kept using strict rules of pedigreed mating, this means retiring both the female and the male. This will make the animals 8-10 months old, an age at which fertility, as measured by litter size, tends to decline for females.[139,140] Over this time period, most inbred strains will have 5-6 litters and most outbred stocks will have 6 litters. Average inbred litter size is 6 for mice and 8 for rats, while the average outbred mouse litter size is 10 and outbred rat litter size is 12. If the mating date of a particular pair or trio is unknown, consider retiring the pair if previous litter size is three or less. Smaller litters are associated with a higher incidence of dystocia, especially in older dams. Some backgrounds and phenotypes (e.g., embryonic lethal or some specific inbred backgrounds such as BN and DBA/2) are associated with smaller litters on average.

Table 7: Developmental milestones of the mouse, day 0-14

Age	Appearance
0 to 24 hours	Deep red Possible milk spot Pigmented mice have dark eyes Umbilicus visible as scab
Day 1	Deep pink Milk spot visible Whiskers more visible Umbilicus visible as scab
Day 2	Ears appear as nubs Milk spot visible Pigment in skin begins to appear Umbilicus visible as scab
Day 3	External ear flap begins to lift from head* Milk spot visible Umbilicus visible as scab
Day 4	External ear flap fully lifted from head and perpendicular to head* Skin fully pigmented Milk spot visible Umbilicus healed
Day 5	External ear flap completely vertical (as opposed to perpendicular)* Skin appears much thicker (milk spot begins to disappear) Incisors visible as white spots under gums
Day 6	Milk spot gone or only faintly visible Colored fuzz appears behind ears or on dorsal neck Incisors erupted
Day 7	Colored fuzz begins to cover pup fully (this is more visible in albino animals, as dark animals may appear "linty" from cage dust)
Day 10	External ear opens Pup fully haired
Day 13 or 14	Eyes begin to open; eye opening is a slit and only becomes oval, then round, over several days after opening

**These events may occur 12-24 hours earlier in CD1 mice (outbred mice appear more developed at birth)*

Table 8: Developmental milestones of the rat, day 0-14

Age	Appearance
0 to 24 hours	Deep red Should see milk spot (if animals have fed) Pigmented rats have dark eyes Umbilicus visible as a scab
Day 1	Deep pink Milk spot visible Whiskers are visible Umbilicus visible as a small scab
Day 2	Ears appear as nubs Milk spot visible Pigment in skin begins to appear Umbilicus still visible as small scab
Day 3	Ears appear as nubs, although some may have begun to lift Milk spot visible Umbilicus visible as small scab
Day 4	External ear flap begins to lift from head Skin fully pigmented Milk spot still visible Umbilicus healed
Day 5	External ear flap fully lifted from head (may be perpendicular to head) Skin appears much thicker (milk spot is more difficult to see; may not be seen in pigmented animals) Incisors visible as white spots under gums Colored fuzz appears behind ears or on dorsal neck
Day 6	External ear flap completely vertical Milk spot gone or only faintly visible, even in albino animals Colored fuzz begins to cover pup fully (this is more visible in albino animals, as dark animals may appear "linty" from cage dust) Incisors erupted
Day 9	Pup fully haired—skin is not visible through haircoat
Day 13 or 14	Eyes begin to open; eye opening is a slit when it first opens, only becoming round over 24-48 hours

PART 3

Genetics for colony management

Genetics is the study of heritable traits and the genes responsible for the variation observed among living organisms. While the field encompasses many types of genetic analysis, from the molecular level to the genetic structure of populations, this chapter will deal primarily with basic Mendelian genetics: the patterns and modes of inheritance from parent to offspring. These principles apply to many naturally occurring traits observed in laboratory populations of mice and rats (coat color, for example) as well as to tracking induced mutations, such as transgenes or knock-out alleles, in engineered populations. Having a basic working knowledge of Mendelian genetics will prove helpful to anyone charged with managing breeding programs of mice and rats.

The start of modern genetics

The observation that offspring inherit traits from their parents had been described long before the establishment of genetics as a science, which is evidenced by selective breeding of plants and animals to the benefit of human populations around the globe. However, the first experiments to try to understand the basis of the observed inheritance of traits did not begin until the mid-19th century.[141] At that time, an Augustinian monk named Gregor Mendel began experimenting in his monastery's garden on the inheritance of several traits exhibited by garden peas. At the time, several theories prevailed regarding the inheritance of traits. One popular theory was the concept of blending inheritance where offspring inherited a smooth continuum of traits from their parents. Another theory was that of the inheritance of acquired characteristics, sometimes referred to as Lamarckian inheritance after its main proponent, Jean-Baptiste Lamarck. This theory held that the experiences of the parents could be directly passed to their offspring. Mendel's work demonstrated that the inheritance of certain traits in pea plants could be predicted and described mathematically. This suggested that heredity was particulate and that the pattern of inheritance of many traits could be explained through simple rules and ratios. It is important to note that terms to describe what he was observing did not exist, so Mendel had no idea what the functional unit of heredity actually was (he referred to these unseen units of heredity as "factors"). We now know that these discrete units are genes, and that variants of a single gene are called alleles.

Mendel was fortunate in his choice of organism to study because peas are a diploid plant species. Diploid organisms have two copies of each chromosome and each chromosome has one allele for a given gene. When an individual has the same two alleles for the gene in question, they are said to be homozygous for that gene (for example, PP or pp). Individuals that carry two different alleles for a gene are heterozygous (Pp). During sexual reproduction, each parent contributes one copy of each chromosome; therefore, the parent contributes one copy of the alleles that parent carries for each gene. One of the characteristics that Mendel studied was the inheritance pattern of pea flower color (purple and white). In the case of flower color, Mendel noted that the flowers were either purple (P) or white (p), but never an intermediate shade. This led to his postulation of the first Law of Heredity, the Principle of Segregation. This law holds that two members of the same factor (gene) segregate, one from the other, into separate gametes (different alleles in each gamete). Mendel's experiments also led to the discovery that alleles of a gene often interact in a specific way, such that the outward appearance (phenotype) controlled by one allele is much more prevalent than that of the other allele. In other words, one allele is dominant to the other allele and its phenotype will be present if an individual has only one copy of that allele. In common genetics convention, dominant alleles are capital letters (P), while recessive alleles are lowercase letters (p).

When Mendel crossed pure-breeding (homozygous) purple-flowered pea plants with pure-breeding white-flowered pea plants this is what he observed:

Parental gametes	P	P
p	Purple offspring (Pp)	Purple offspring (Pp)
p	Purple offspring (Pp)	Purple offspring (Pp)

When he then intercrossed the offspring produced above, this is what he observed:

Parental gametes	P	p
P	Purple offspring (PP)	Purple offspring (Pp)
p	Purple offspring (Pp)	White offspring (pp)

So, for the gene that controls pea flower color, the allele for purple color (P) is dominant to the allele for white color (p), thus this allele is called recessive. The charts used above to show how the gametes segregate and then come together in the offspring are called Punnett squares, and are used by geneticists to illustrate the results of crosses between individuals of known genotype(s). Several other genetic terms are introduced with the above examples. The first cross is called a "parental cross" and is designated as generation P1. The offspring produced from that cross are the "first filial" (F1) generation. Crossing the F1 offspring produces F2 (second filial) offspring in a 3:1 phenotypic ratio (for this example) of purple:white flowers. The genotypic ratio is 1:2:1 and these ratios held for many of the single gene traits that Mendel studied in peas. These ratios are always present in heterozygous crosses involving a single gene with two alleles that are dominant or recessive.

In addition to flower color, Mendel also studied pea color and shape; yellow/green and round/wrinkled, respectively. Since both of these traits are found within the same "individual" (the pea), results from crosses studying the inheritance of these characteristics led Mendel to propose his second Law of Heredity, the Principle of Independent Assortment. This law holds that segregation of one trait is independent of segregation of the other trait. As with the crosses involving flower color, Mendel started by intercrossing pure-breeding green (YY) and smooth (RR) parental plants with yellow (yy) and wrinkled (rr) plants which produced F1 offspring that were all green and smooth for phenotype (Rr Yy), indicating that the alleles for green and smooth were dominant to those for yellow and wrinkled. When the F1 offspring were intercrossed, Mendel observed the following:

Parental gametes	YR	Yr	yR	yr
YR	Green, Smooth (YYRR)	Green, Smooth (YYRr)	Green, Smooth (YyRR)	Green, Smooth (YyRr)
Yr	Green, Smooth (YYRr)	Green, Wrinkled (YYrr)	Green, Smooth (YyRr)	Green, Wrinkled (Yyrr)
yR	Green, Smooth (YyRR)	Green, Smooth (YyRr)	Yellow, Smooth (yyRR)	Yellow, Smooth (yyRr)
yr	Green, Smooth (YyRr)	Green, Wrinkled (Yyrr)	Yellow, Smooth (yyRr)	Yellow, Wrinkled (yyrr)

This type of "bigenic" or "dihybrid" cross, where two traits are assorting independently, produces many more phenotypic classes:

- 9 green/smooth
- 3 yellow/smooth
- 3 green/wrinkled
- 1 yellow/wrinkled

Nine different genotypes underly the visible phenotypes. Although they can become quite complex, Punnett squares are a useful tool in working out the combinations of alleles that produce phenotypes and genotypes for any number of traits that are controlled by independent genes.

Modes of inheritance

Punnett squares are the tool of choice for experimental genetics, but scientists interested in studying the inheritance of human disease often use a different tool called a pedigree chart. These diagrams are used to track the occurrence, by sex, of a disease phenotype in all relatives of a family. Since the inheritance pattern is being tracked by sex across generations, pedigree analysis can be used to determine the underlying genetic control of a novel phenotype. Of particular interest are the modes of inheritance. These include the type of allele (dominant or recessive) responsible for the phenotype, and whether the gene is on an autosome or sex chromosome (X or Y). An autosome is simply a chromosome that is not a sex chromosome, and males and females of a species have equal numbers of autosomes, although the number of autosomes differs across species. For example, mice have 20 pairs of chromosomes; 19 autosomal pairs and a pair of sex chromosomes while rats have 21 chromosomal pairs (20 autosomal pairs and a pair of sex chromosomes). *Figure 18* gives some examples of pedigree charts illustrating autosomal dominant and recessive modes of inheritance, as well as examples of sex-linked inheritance.

Figure 18. Pedigrees illustrating various patterns of inheritance for single gene autosomal and sex-linked traits. Recessive traits are often described as "skipping a generation." This phenomenon may be seen in this illustration.

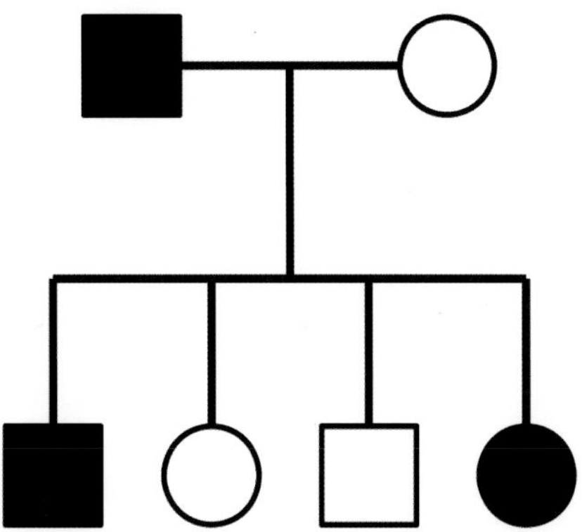

A) Autosomal dominant; male is a heterozygous carrier. There is a 50% chance of producing affected offspring of either sex.

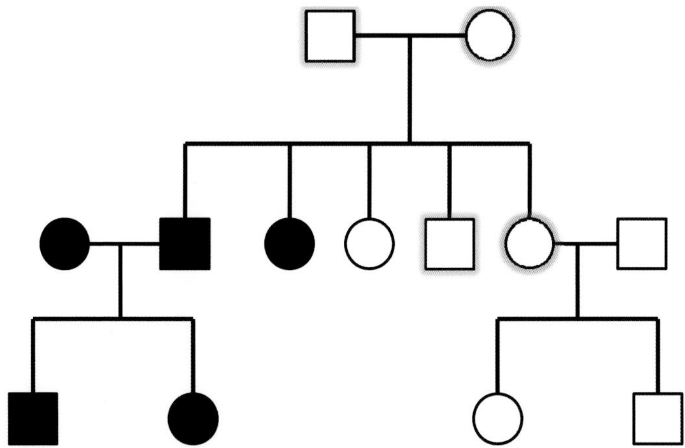

B) Autosomal recessive; both first-generation male and female are carriers. The mutation is difficult to track unless there is a test for carriers or offspring are homozygous. The mutation can be eliminated from branches in a pedigree (2nd generation on the right; all offspring are unaffected).

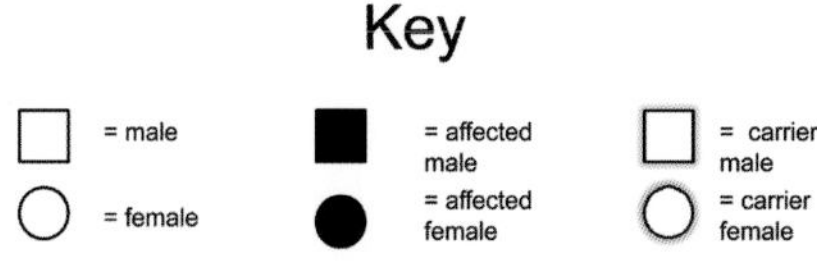

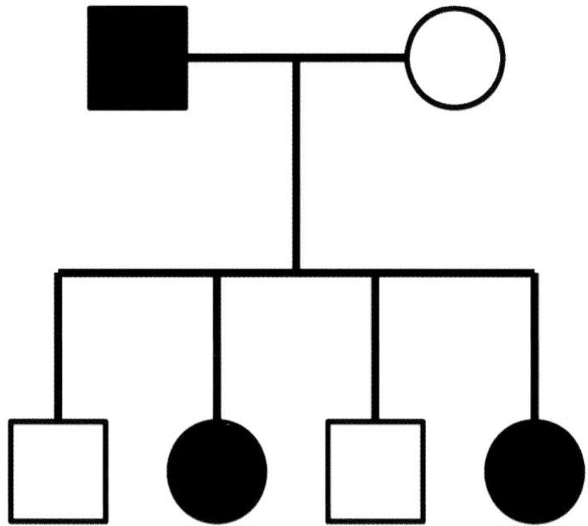

C) X-linked dominant; parental male is a carrier. All daughters of the affected male parent will be affected, but no sons will inherit the mutation.

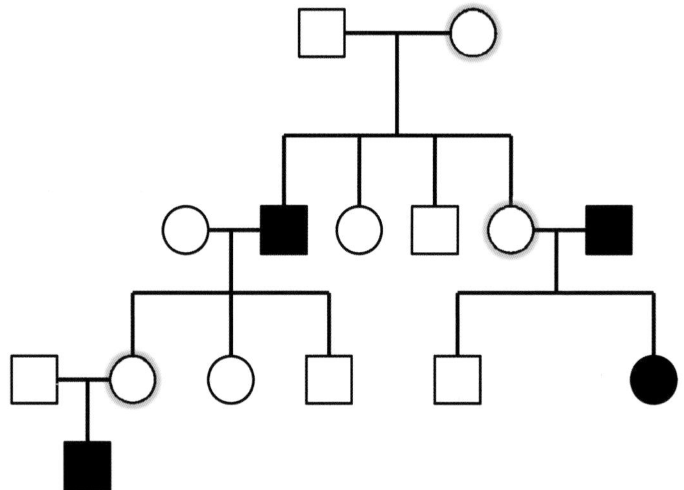

D) X-linked recessive; parental female is heterozygous carrier. Sons of a carrier female have a 50% chance of inheriting the mutation and thus being affected. Daughters of a carrier female have a 50% chance of being carriers, but none will be affected.

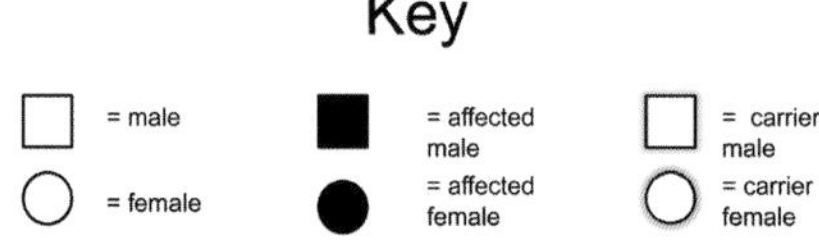

Practical genetics for the mouse room

Understanding Mendelian genetics and modes of inheritance provides a good foundation for understanding several aspects of day to day genetics in a working mouse room. For the most part, animal room populations are composed of "wild-type" mice and/or genetically engineered mice. The wild-type mice are typically vendor-derived, inbred strains that have various characteristics better described elsewhere in this manual. However, all of the inbred strains in common use have the phenotype of coat color that can be used as a tool for genetic management (in particular quality control) of a colony. There are entire books written on the coat colors of mice and the genes and alleles involved,[142] but for most considerations there are four primary coat color genes that influence the variation seen in common inbred strains:

1. Agouti vs non-agouti, commonly denoted by A and a, respectively.
2. Black vs brown, classically denoted B and b, respectively, but the gene is now known and the alleles are *Tyrp1*$^{+}$ or *Tyrp1*^{b}.
3. Pigmented vs albino, classically referred to as C and c, respectively. This gene is also known and the most common alleles are *Tyr*$^{+}$ or *Tyr*^{c}.
4. Non-dilute vs dilute, classically noted as D and d, respectively. This gene has also been identified and the alleles are *Myo5a*$^{+}$ or *Myo5a*^{d}.

As is typical with genetic nomenclature, the capitalization of the letters used to denote these coat color alleles indicate which allele is dominant (A, for example) and which is recessive (a). Since inbred strains are homozygous at all loci, it can be inferred that strains of a particular coat color have a specific allele at the locus that controls that color. For example, any agouti inbred strain would be AA at the agouti locus (Agouti 129 strains carry a variant of A, A^{W}, which results in a light belly). However, since the deposition of pigment is an inter-related pathway, there are interaction effects across these four loci that control steps in the pathway and the ultimate coat color generated. Animals that are AA, but are also BB (normally resulting in a black coat color) will be agouti since that locus will directly influence the deposition of black pigment. The agouti C3H/HeNCrl is an example of this with the coat color alleles AABBCCDD. Animals that are aa and BB will be black, as the striping pattern of the agouti hair is absent. The classic example of this is the C57BL/6NCrl, which is aaBBCCDD. Another example is the "d" allele which influences the amount of pigment produced. The strain DBA/2NCrl has aabbCCdd for the major coat color alleles so it has a very light brown coat color when compared to a strain that is aabbCCDD. Albino strains of mice lack any pigment due to mutations in the C (*Tyr*) gene, regardless of what other coat color alleles they carry. The albino locus is said to be epistatic to other coat color loci since it is unlinked and completely masks their phenotypic expression. BALB/cAnNCrl and FVB/NCrl are two common albino strains with different allelic combinations masked by homozygosity for a mutant form of the C locus. BALB/cAnNCrl are AAbbccDD while FVB/NCrl are AABBccDD. Knowledge of the coat color alleles present in strains in a colony *(Table 9)* can help determine the source of genetic contaminations if unexpected coat colors suddenly appear in breeding cages.

Table 9: Coat color alleles of commonly used inbred strains of mice.

Strain	**Agouti (A or a)**	**Non-agouti (B or b)**	**Albino (C or c)**	**Dilute (D or d)**	**Coat color**
129S2/SvPasCrl	A^WA^W	BB	CC	DD	Light-bellied agouti
BALB/cAnNCrl	AA	bb	cc	DD	Albino
C3H/HeNCrl	AA	BB	CC	DD	Agouti
C57BL/6NCrl	aa	BB	CC	DD	Black
DBA/2NCrl	aa	bb	CC	dd	Light brown
FVB/NCrl	AA	BB	cc	DD	Albino

While it is good to have an understanding of basic coat color genes and their interactions from a quality control standpoint, one also needs to apply basic genetics to tracking mutations that are being maintained in a colony. In particular, this is required if the colonies are producing animals that have been genetically engineered, and investigators require specific genotypes for their experiments. The methods used for producing cohorts of animals in particular numbers with particular genotypes are detailed elsewhere in this guidebook, so this section will briefly examine the genetics behind those calculations. Note also that we are not concerned with whether an allele is dominant or recessive (although one may need to consider if the mutation is lethal when homozygous, for example), so the main considerations for determining the genetic outcome of specific crosses are:

1. Genotype of the parents in the cross for the mutation(s) in question.
    a. Heterozygous
    b. Homozygous
    c. Wild-type
2. Location of the mutation(s) in question.
    a. Autosomal
    b. Sex-linked (almost always X chromosome)

Once the above information is known, it is relatively straightforward to calculate the probability of producing specific genotypes in the offspring. Punnett squares are an excellent tool for this purpose, particularly if more than one manipulation is present in the animals being mated, and if particular combinations of mutations are desired in the offspring. Two of the most common crosses are illustrated

1. Heterozygote (Mut/+) by wild-type (+/+) where Mut denotes the manipulation and + denotes the normal allele.

Gametes	Mut	+
+	Mut/+	+ / +
+	Mut/+	+ / +

Offspring are 50% heterozygous and 50% wild-type.

2. Heterozygote (Mut/+) by heterozygote (Mut/+).

Gametes	Mut	+
Mut	Mut/Mut	Mut/+
+	Mut / +	+ / +

Offspring are 25% homozygous for the Mut allele, 50% heterozygous, and 25% homozygous for the wild-type allele.

below since they form the basis for calculating genotypic frequencies for more complicated crosses involving two or more manipulations:

If the cross involves parents that are carrying two different unlinked (on separate chromosomes) manipulations, the probabilities for various combinations of the two mutations in a single offspring can be calculated with or without Punnett square diagrams as long as the zygosity of the parents for each mutation is known. Taking example #2 above, but with each parent being double heterozygotes for the same pair of unlinked mutations, the calculation for a given genotypic combination in a single offspring is the product of the two individual probabilities for a particular combination. For example, the probability for either mutation being homozygous in an individual is 25% for each mutation. So, the probability of both mutations being homozygous in the same individual is 0.25 x 0.25 = 0.0625 or about 6%. Likewise the probability of a given offspring being homozygous for one mutation and heterozygous for the other is 0.25 x 0.5 = 0.125 or about 12.5%. The probability of any combination possible can be calculated in this manner or can be visualized in a Punnett square like that of the bi-genic cross shown earlier (the double homozygote, YYRR, has a frequency of 1/16 or 0.0625).

Predicting genetic outcomes from crosses involving sex-linked mutations is very similar to the examples given above.

Figure 19 . Inheritance patterns of a sex-linked mutation are determined based on both the carrier status of parents and the sex of resulting offspring. Since males have one X chromosome, they either have the mutant allele (hemizygous) or not. Females, with two X chromosomes, may be homozygous, heterozygous, or may have no copies of the allele.

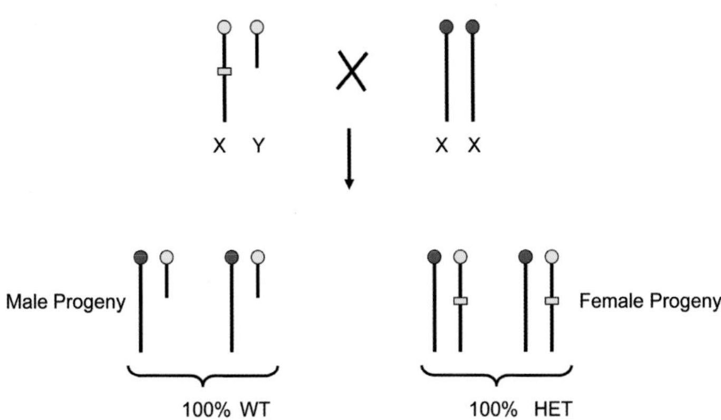

Figure19 A) A cross of an hemizygous male with a wild-type female.

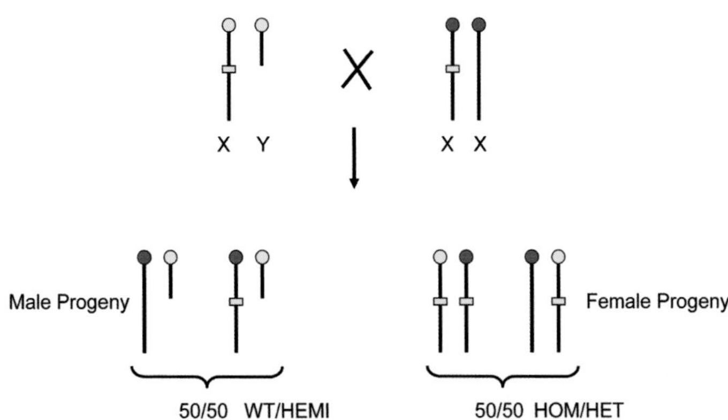

Figure19 B) A cross of a hemizygous male with a female heterozygote.

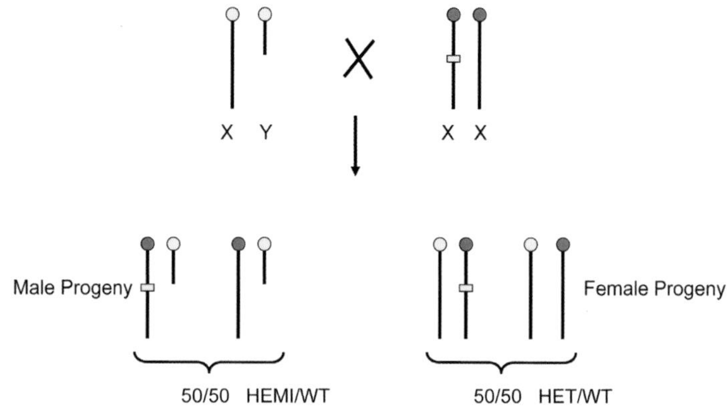

Figure19 C) A cross of a wild-type male with a heterozygous female.

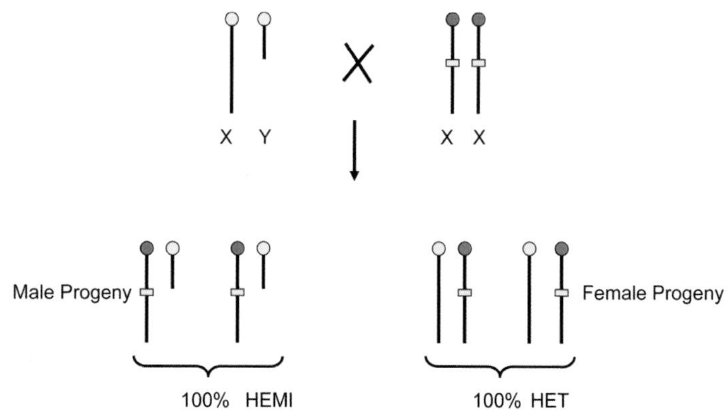

Figure19 D). Cross of a wild-type male with a homozygous female.

However, the probability of a specific genetic combination will also depend upon the genotypes of the male and female parents, as well as the sex of the offspring produced. *Figure 19* illustrates examples of sex-linked inheritance.

These illustrations of X-linked inheritance and probabilities also introduce a new genetic term, hemizygote. Hemizygosity is the state of having unpaired regions on a chromosomal pair. Since the X and Y are considered a chromosomal pair, genes on the X or Y do not have paired regions in male mice since they only have one copy of each sex chromosome. This condition can also be present on the autosomes of random integration transgenic animals, and this type of manipulation will be discussed in greater detail in the next chapter.

PART 4 Transgenic technologies

The study of mutant mice has evolved from collections of spontaneous coat color mutants held by 19th century mouse fanciers, to the advent of directed manipulation of the mouse genome by several methods. Today, there are repositories of genetically engineered mice located around the world that provide scientists with access to many disease models. These mice have been engineered using techniques like targeted mutagenesis, inducible mutagenesis, and transgenesis. This chapter will provide an introduction to these technical procedures and discuss the advantages and disadvantages to each approach. There are several manuals, such as those by Nagy, Pinkert, Cartwright, Kühn and Wurst, and Joyner that describe these procedures in great detail, and are an excellent reference for those interested.[143-147]

Transgenesis

Transgenesis involves introduction of known genes into the mouse genome at random sites with the intent to produce a phenotype based on overexpression of the gene. This method was first used in the early 1980s[148] and led the way for other more directed methods that followed. Transgenes can be completely "assembled" in the laboratory by linking together various components that will allow for the cloning (the vector), the expression (the promoter), and the processing and protein coding (intron, exons/cDNA, and polyandenylation signal) of the transgene once it is introduced into the mouse genome. The cDNA is DNA that is synthesized from messenger RNA (mRNA) using an enzyme called reverse transcriptase. The mRNA is a copy of the regions of a gene, the exons, that code for the protein that the gene ultimately produces. In the genome, genes are made up of multiple exonic sequences with intervening non-coding regions, the introns, between them. Promoters are regions of DNA that drive and control gene expression in the cell. They can allow for ubiquitous expression of a gene in all cell types (sometimes called a constitutive promoter), they can only allow expression within a specific cell or tissue type (liver-specific, heart-specific, neuron-specific, etc.), and they can be made to respond to factors that turn on expression at a specific time (inducible promoters).[149] Regardless of the type of promoter, these regions in the construct allow for expression of the transgene independent of genomic location. *(Figure 20)* If a transgene is assembled in this manner, protein may be expressed in tissues it would not normally occupy and at levels that are higher than normal. These types of changes often lead to phenotypes that can provide clues about the mode of action of specific proteins, or serve as models for disease. It is not unusual for the protein coding region to come from species other than the mouse (human cDNA sequence is often used). Transgenes can also be made by cloning large regions of genomic DNA that contain the gene of interest complete with its natural controlling elements (promoter) and exon-intron assembly intact. Although these larger constructs are a technical challenge, they may allow for better expression of the gene of interest for reasons explained below.[150]

Transgenes of either sort are introduced to the mouse genome via pronuclear microinjection of mouse one cell embryos. Embryos are collected from females early in the day following mating by flushing them from the reproductive tract into tissue culture media. They are then placed on a microscope stage and specialized pipettes are used to hold the embryo or inject

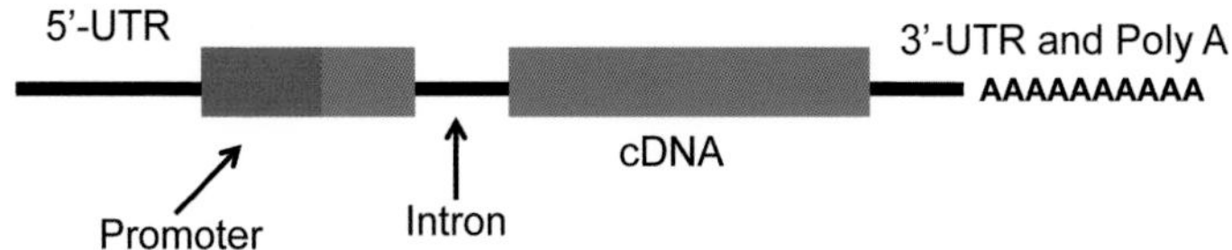

Figure 20. Schematic diagram of a typical vector for the production of transgenic mice. 5' and 3' untranslated regions (UTR) can contain elements that promote expression of the transgene after insertion in the host genome.

the DNA. *(Figure 21)* Surviving embryos are then transplanted into pseudopregnant females and carried to term. Once the litter is born, pups are assayed for the presence of the transgene in their genome. This method typically results in random integration into the genome at only one site per individual (called a transgenic line founder animal), but this integration site may contain a tandem array of transgenes linked together in multiples ranging from two to several hundred. Multiple founder animals may be produced following an injection session with a given transgene and although they contain the same transgene, it is important to note that each founder animal is unique. This is due to the fact that the site of integration and the size of the tandem array will be different for each one. Those differences often lead to different levels of expression of the transgene and, therefore, to potentially different phenotypes in offspring of the founder animal. After the founder animals are produced, the next step is to breed them to wild-type animals to see if the transgene is passed on to offspring. If germline transmission of the transgene occurs, carrier animals can be used to test for level of expression of

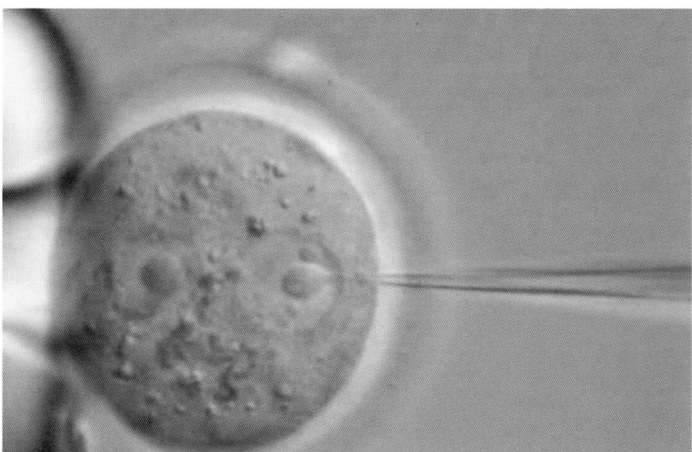

Figure 21. Bright-field photo of a one-cell mouse embryo being microinjected. The rounded pipette on the left is a holding pipette that stabilizes the embryo for insertion of the injection pipette on the right. The injection pipette contains the linearized transgene vector that is being injected into the pronucleus of the embryo.

the transgene, and founder lines that produce animals with suitable amounts of transgenic protein production can be used for further study.

Transgenesis methods can also be used to produce mice that have specific utilities when combined with some of the other types of manipulation that will be discussed later. This is particularly the case for transgenes that have inducible promoters. These types of promoters are controlled by an external stimulus that activates them to start driving expression of the cDNA they are connected to. The most widely used is a binary transcription transactivation system called the "tet-on/tet-off" system.[151] The transcriptional activator gene can be regulated reversibly and quantitatively by the antibiotic tetracycline or a derivative like doxycycline (dox). Doxycycline can be delivered to the transgenic animals via drinking water or chow at a prescribed time to activate (or repress) expression of the transactivator protein. As illustrated in *Figure 22*, the tet-on system activates transcription in the presence of dox, while the tet-off system represses transcription in the presence of dox. Requiring fairly high doses of antibiotic over a long time is the major drawback to these systems. Tet-off requires continuous administration of dox, and activation only occurs once dox is cleared, which can take anywhere from 1-7 days in adult animals. In addition, dox is cleared more slowly in tissues like bone and liver. The tet-on system induces transcription rapidly, but repression depends on the clearing of dox.

Transgenesis performed in this manner has two main difficulties. The first is that the random integration of the transgene may result in the disruption of an endogenous gene. If the disruption is enough to ablate or change expression of the gene, a phenotype other than the one desired may result. In a worst-case scenario, researchers may attribute an observed change in their transgenic line to the transgene itself, when it is really due to disruption of an unrelated gene. The second difficulty is with transgene expression levels that can vary considerably from founder line to founder line for the same construct. Expression levels are

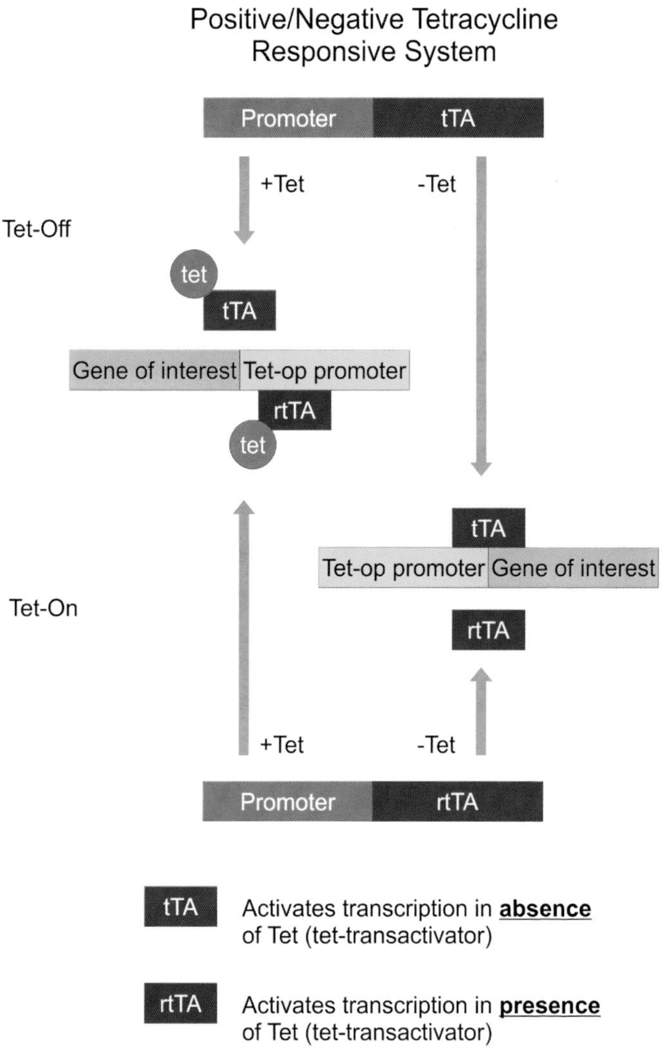

Positive/Negative Tetracycline Responsive System

Promoter tTA

+Tet

-Tet

Tet-Off

tet

tTA

Gene of interest Tet-op promoter

rtTA

tet

tTA

Tet-op promoter Gene of interest

Tet-On

rtTA

+Tet

-Tet

Promoter rtTA

tTA Activates transcription in **absence** of Tet (tet-transactivator)

rtTA Activates transcription in **presence** of Tet (tet-transactivator)

Figure 22. Diagram of the the function of tet-on and tet-off approaches for the regulation of transgene expression.

greatly affected by the genomic insertion site of the transgene, with heterochromatic regions (areas of the genome that have few or no active genes) tending to silence transgene expression altogether.[152] One reason for using large, genomic DNA-based transgenes is to insulate the expression controlling regions (promoter) from the surrounding genomic region.

Targeted mutagenesis

As the name implies, targeted mutagenesis is the purposeful manipulation of a specific gene target (or genomic region) to produce a desired effect. The two most common types of targeted mutagenesis are knockouts and knockins.[153] Knockouts are designed to eliminate gene function and can be either constitutive (gene function is eliminated in all tissues) or conditional (gene function is eliminated in an inducible or tissue-specific manner).[154,155] Knockins are produced by targeting a construct, for example, a transgene, to a specific location in the genome. They too can be constitutive or conditional in terms of gene expression control. Details about the advantages and disadvantages of a constitutive versus conditional approach will be covered later in this chapter. For both knockouts and knockins, targeting of the gene or genomic region is carried out in embryonic stem (ES) cells using vectors that are capable of "finding" the region they are specifically meant to target. Embryonic stem cells are derived from the inner cell mass (ICM) of 4.5-5 day old mouse embryos (blastocysts). Blastocysts consist of a hollow ball of cells that will become both the embryo and its supporting tissue (placenta) once the embryo embeds in the uterine wall. The ICM is contained within this ball and is composed of ES cells that are capable of becoming any tissue in the developing mouse (they are pluripotent).[156-158] Mouse ES cells were first derived from the 129 inbred strain, but have since been cultured from other inbred lines (e.g., C57BL/6) and a number of different cell lines are currently available to the scientific community.[159,160]

The production of knockout animals begins with the selection of the target and the isolation (cloning) of a portion of the genomic copy of the gene. That section becomes the basis for building a targeting vector that will ultimately be introduced into ES cells. Targeting vectors can be quite complicated in their design, but the basic vectors are intended to disrupt a portion of the normal coding region of a gene. They also contain a selectable marker such as neomycin (Neo) that helps select for ES cells that have taken up the targeting vector, hopefully at the appropriate site. They may also contain a negative selection cassette such as thymidine kinase (TK)[161,162] to help eliminate cells where the targeting vector has integrated at a random site. *Figure 23* is a diagram of a basic targeting vector containing the Neo positive selection cassette along with a TK negative selection cassette compared to the wild-type allele.[163] For correct targeting to occur, the targeting vector must recombine with the wild-type locus via the process of homologous recombination. Homologous recombination involves the physical exchange (sometimes called crossing-over) of DNA between two regions of similar or identical nucleotide sequence.[164] In this way, the mutated region from the targeting vector is "swapped" for the corresponding region in the wild-type allele resulting in a disruption of gene function as illustrated in *Figure 24*. Cells that are correctly targeted will have the Neo cassette integrated, but the TK cassette will be lost since it is outside of the region of homology between the targeting vector and endogenous locus. Subsequent selection with neomycin in culture will kill cells that do not have a neomycin cassette and enrich those cells that do. However, the targeting vector may integrate into the ES cell genome at random (as

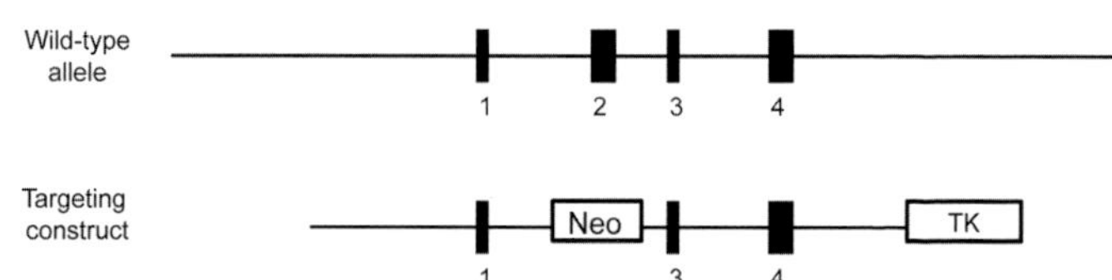

Figure 23: Diagram of a basic targeting vector, used to replace a gene by homologous recombination, and the complementary wild-type allele for comparison.

described above for transgenes) and these cells will survive Neo selection in culture as well. Use of the TK vector helps reduce the total number of surviving cells since many of the cells that have random integrations will also have an intact TK cassette. Selection with agents that kill thymidine kinase positive cells will eliminate the cells with random integrations, further enriching correctly targeted cells. Since use of the positive and negative selection system does not work as planned 100% of the time, some cells will survive and grow even though they are not correctly targeted. The selection methods just reduce the total number of ES cell clones that need to be examined for correct targeting. Given that targeting efficiencies can be lower than 1% for certain regions, positive-negative selection can go a long way toward reducing the total number of ES clones that have to be examined for that rare homologous event.[165]

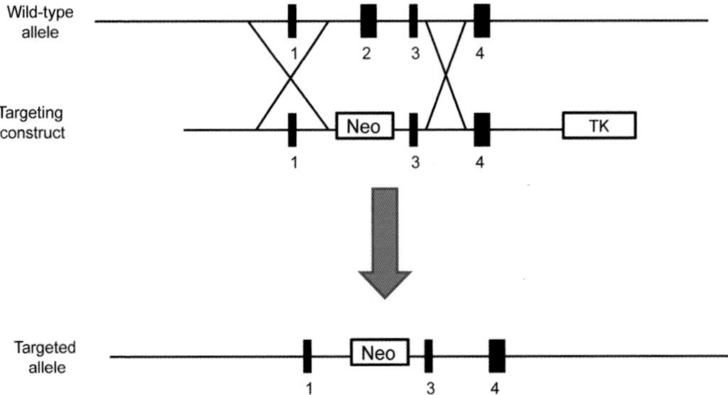

Figure 24. Diagram of targeting vector replacement of an endogenous locus via homologous recombination. Homologous recombination between the targeting vector and the endogenous allele results in the introduction of the Neo cassette into the locus and the deletion of the TK cassette. This allows for screening of ES cell clones for homologous recombination by the addition of G418, an antibiotic related to gentamycin, to the cell culture medium. Incorrectly targeted clones retain the TK cassette and can be eliminated through the addition of acyclovir or gancyclovir to the culture medium. Use of both positive and negative selection methods greatly reduces the number of clones that must be screened using more labor intensive methods such as Southern blotting.

Once correctly targeted cells are identified, they are expanded and then prepared for injection into host blastocysts.

As shown in *Figure 25*, this process is analogous to that described for random integration transgenesis, with only the stage of the embryo (blastocyst versus one-cell), the size of the injection pipette (large-bore versus small-bore needle) and the material being injected (ES cells versus DNA) being the major differences. After injection, the surviving blastocysts are transplanted into pseudopregnant recipient females; however, the resulting offspring may not be wholly derived from a single cellular origin as they are in random integration transgenesis. The reason is that only a small number of targeted ES cells are injected into the host blastocyst. Blastocysts contain a large number of wild-type (host-derived) ES cells in the ICM, and the hope is that the injected ES cells will be incorporated as various tissue types in the developing embryo. If that occurs, the resulting animals will be chimeric; some tissues will be derived from the host ES cells and some from the targeted ES cells. Those tissues derived from injected cells will also contain the mutation, so it is essential for germinal tissue (cells that will become sperm in most cases) to be

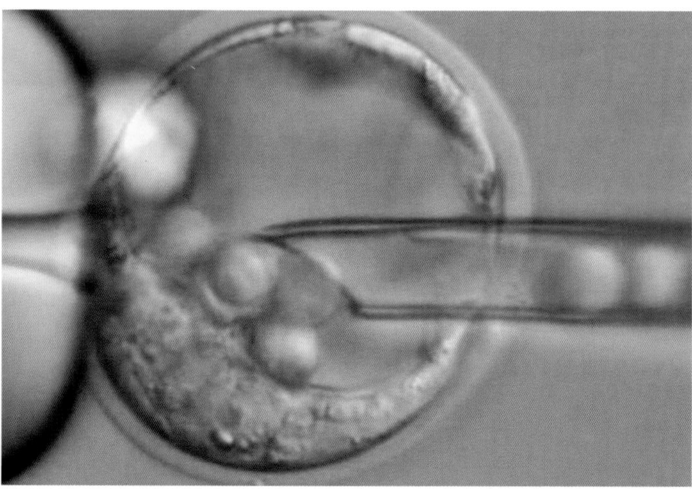

Figure 25. Bright-field photo of blastocyst injection with manipulated ES cells. The rounded pipette on the left is a holding pipette that stabilizes the embryo for insertion of the injection pipette on the right.

derived from the targeted ES cells if the mutation is to be passed on to the next generation. *Figure 26* is a diagram of the overall process to produce gene-targeted mice.

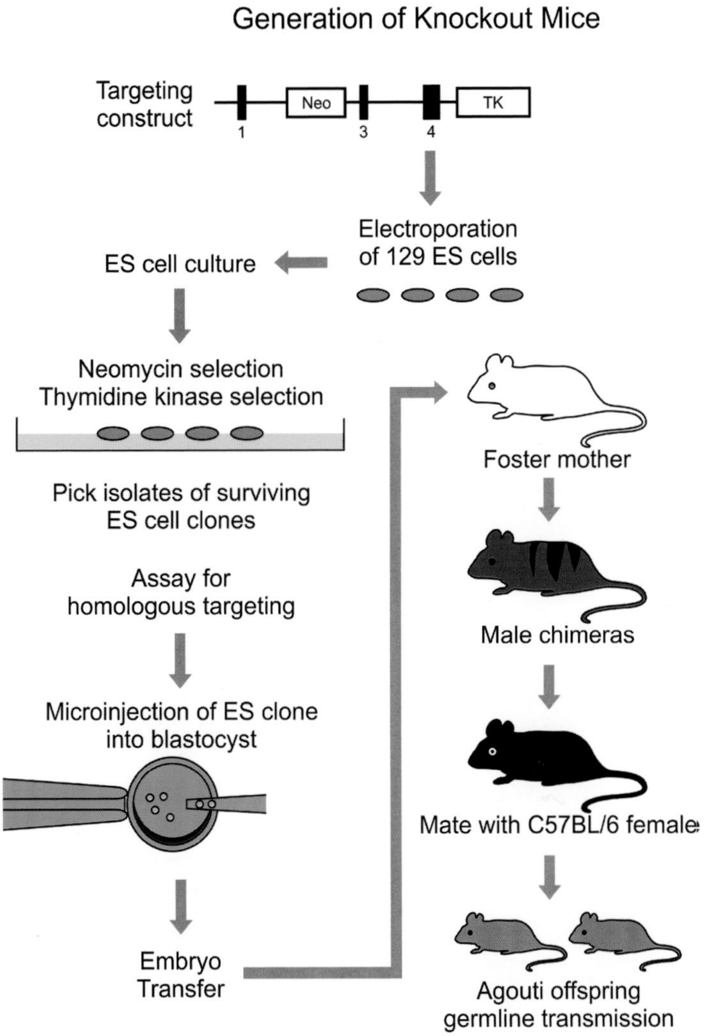

Generation of Knockout Mice

Targeting construct
Neo
TK
1
3
4
Electroporation of 129 ES cells
ES cell culture
Neomycin selection
Thymidine kinase selection
Pick isolates of surviving ES cell clones
Assay for homologous targeting
Microinjection of ES clone into blastocyst
Embryo Transfer
Foster mother
Male chimeras
Mate with C57BL/6 female
Agouti offspring germline transmission

Figure 26. Overview of the processes involved in targeted mouse model creation. From targeting vector construction to F1 knockout animals usually takes 6-8 months.

The final two steps introduce another "trick" that is used by scientists to help identify animals that are high-percentage chimeras (most of the tissues being derived from injected ES cells) through visible inspection. If the host blastocyst is selected from a strain that has different coat color alleles than those of the ES cell strain origin, the resulting chimeras will show a mix of the two coat colors because some hair cells are derived from each strain. A high percentage of the coat color expected for the injected ES cells is a good indication that much of the other tissues were derived from mutant ES cells. These high-percentage chimeras can then be bred to wild-type animals to test for germline transmission of the mutant allele. Once again, animals that are derived from targeted ES cell sperm origin can be identified in litters based on coat color if thought is given to the various coat color alleles involved. For example, ES cells derived from an agouti 129 inbred strain will be A^WA^WBBCCDD (see Table 9 for coat color allele information) and produce sperm that is A^WBCD. If the host blastocysts are derived from the black C57BL/6 inbred strain (aaBBCCDD), sperm produced from host ES cells will be aBCD. If chimeras composed of these two strains are mated with C57BL/6 animals the resulting offspring can be either black (aaBBCCDD) or agouti (A^WaBBCCDD). Any agouti animal will have been produced by sperm from the manipulated ES cell and will have a 50% chance of carrying the mutation, so one only needs to screen agouti animals to identify carriers. Multiple litters of all black offspring indicate that the manipulated ES cells did not contribute to the germline (sometimes this happens even in high-percentage coat color chimeras) and more chimeras will need to be made and bred. Once germline transmission is confirmed, animals can be bred in the usual fashion to expand the colony and produce genotypes (mutant homozygotes in particular) for further study of the expected phenotype.

Inducible mutagenesis

Early in the development of targeted mutants, most if not all of the vectors were designed to produce null (loss of gene function) mutations. In some cases, genes that were known to produce a particular phenotype, such as a human disease, were targeted in an attempt to make a mouse model to study the disease. In other cases, genes of unknown but suspected function were targeted and the resulting animals were examined for phenotypes of interest. While both are seemingly straightforward approaches, it was soon realized that many targeted genes had no obvious phenotype, produced a phenotype that was unexpected, or were lethal. Since these were null mutants, all cells in the body contained the mutation thus the gene would not function regardless of tissue type. This type of manipulation is called a constitutive mutation because the knockout is always "on." If the gene in question is necessary for embryonic development, no live animals will be produced making subsequent analysis difficult at best. To avoid this problem, scientists developed a system that would allow for the production of targeted alleles. Those alleles function normally until a special signal was given to create a null allele. These types of models are called conditional (or inducible) knockouts since the disruption of the gene only occurs under certain conditions. *Table 10* lists the characteristics of constitutive and conditional approaches to targeted mutagenesis. The last row of that table introduces the primary mechanism used to manipulate conditional alleles, the use of recombinase mediated gene editing.

Recombinases are a class of enzymes capable of rearranging sections of DNA if specific recognition sequences are present in the genome. The most widely used of these for engineering targeted alleles in the mouse is cre recombinase from P1 bacteriophage.[166] This enzyme catalyzes site-specific recombination of DNA between sequences of DNA called "loxP" sites so that the sequence between two loxP sites is

Table 10. Constitutive versus conditional strategies

Constitutive	Conditional
ES cell mutated	ES cell wild type
KO/KI present in all cells	Cell-specific
Promoter independent	Promoter dependent
No regulation of mutation	Mutation can be regulated
No recombinase transgenic line required	Recombinase transgenic line required

removed. Since these sequences are not normally found in mammalian genomes, they can be artificially introduced into targeting vectors to flank regions that are intended to be removed later. With clever design, their presence will not perturb normal gene function so the targeted locus will behave as a wild-type locus until the cre recombinase is present. In addition to the cre-loxP system, the analogous flp-frt recombinase system originally isolated from yeast can also be added to targeting constructs.[167] When present, flp recombinase will find its frt recognition sites and remove DNA regions that lie between them. Combining the cre-loxP with flp-frt in the same targeting vector has allowed for the construction of a very complex, targeted allele, where regions of the mouse genome can be rearranged in a stepwise manner to create multiple knockout alleles.

Further refinement to targeted allele generation in mice is accomplished by using random integration transgenic lines that have cre or flp recombinase expression under the control of inducible and/or tissue specific promoters.[168] Presently there are many transgenic lines available that have recombinase expression controlled by promoters, each active in a specific tissue or cell type that allows spatial control of targeted allele generation.[169,170] In addition, lines that have inducible promoters driving recombinase expression can be employed for temporal control of knockout alleles. Crossing these transgenic lines with mice that have recombinase recognition sites engineered into a gene allows for the production of animals with disruptions in genes that would otherwise cause embryonic lethality in a constitutive knockout.[171] *Figure 27* is a schematic of a system where cre expression is driven by a tissue specific promoter so that the excision of the targeted locus occurs only in cells where the promoter is active. Post-recombination, a single loxP site remains in the endogenous locus, while the region that was flanked by loxP sites is circularized and lost.

Taken together, the methods described above have allowed for the production of mouse models of human disease that are extremely refined and useful for the study of potential therapies. Although the mouse is not a perfect model organism for some human conditions, recent advances in gene targeting in the rat will allow for the production of better models for certain diseases.[3,4,172,173]

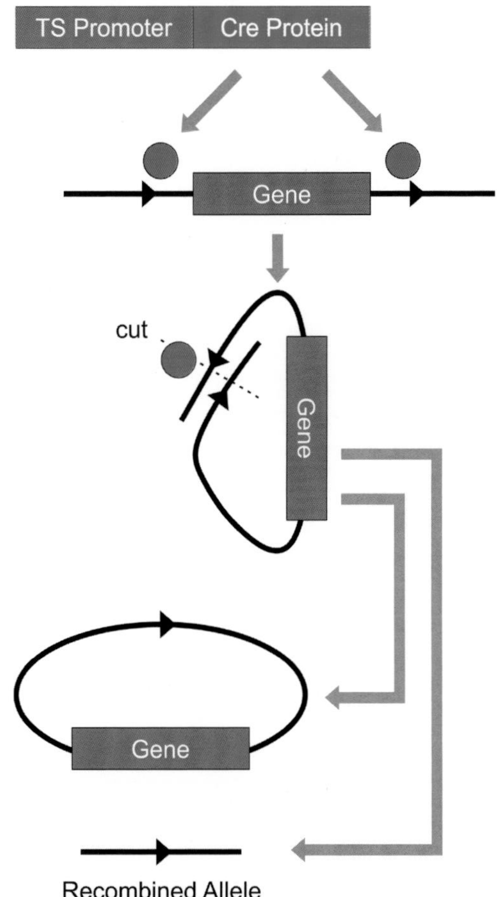

Figure 27. Diagram of cre mediated excision of a loxP flanked region. The cre protein is produced by a tissue-specific promoter. It acts on the loxP sites, first circularizing, then excising them, leaving a circular portion of DNA containing the flanked gene and the recombined genomic DNA.

PART 5 Laboratory mouse and rat nomenclature

There are four rules to remember when deciphering the nomenclature of mice and rats. First, look for the name of the animal. Second, look for the gene and/or allele. Third, if it can happen to or be performed on a rodent genome, there is a formal rule of nomenclature to address the genomic change. And finally, always remeber these rules can change. The most current information on mouse and rat nomenclature is always available at http://www.informatics.jax.org/mgihome/nomen/index.shtml and http://rgd.mcw.edu/nomen/rules-for-nomen.shtml. On both sites, information on submitting new animal, gene, or allele names may also be found.

As new and different ways to manipulate the mouse and rat genome are developed, new mouse and rat variants are created. Each of those variants, whether the variation was spontaneous or induced, has a standard nomenclature to clearly communicate what has happened to that animal's genome. Mouse and rat animal and gene nomenclature follows the rules and guidelines established by the International Committee on Standardized Genetic Nomenclature for Mice and the Rat Genome Nomenclature Committee. As new technologies are developed, these must be accommodated by nomenclature, so the rules and examples below are only the most basic, and do not include an exhaustive list of all nomenclature rules for all possible manipulations.

The scientific classification of mice and rats is the same to the genus level: order Rodentia (from the Latin *rodere*; to gnaw), superfamily Muroidea; family Muridae; subfamily: Murinae. The scientific name of the domestic laboratory rat is *Rattus norvegicus*, while the proper scientific name of the laboratory mouse is *Mus musculus musculus* x *domesticus*, or the laboratory mouse, indicating their status as hybrids of two subspecies.[174] *Mus* is derived from the Sanskrit; *mush*, to steal, while the origin of *Rattus* is unclear (possibly post-classical Latin). Other wild-derived *Mus* species mice may be seen in use in the laboratory setting: *M. musculus castaneous* (from Thailand), *M. m. molossinus* (from Japan), and *Mus spretus* (from North Africa).

There are three basic types of mice and rats in use in research today: wild-caught or wild-derived animals, outbred stocks, and inbred strains. Wild mice and rats are generally just called by their scientific name. Care should be taken to use correct binomial nomenclature (standard scientific nomenclature; *Genus species*) and current species names. For example, *Peromyscus maniculatus*, or *Rattus rattus* may be used to refer to deer mice or black rats, respectively.

Before venturing further into nomenclature, laboratory codes must be addressed since they are seen in every subsequent type of nomenclature explained here. Laboratory codes are unique 1-5 letter identifiers that individuals, laboratories, or institutions can register with the Institute for Laboratory Animal Research (ILAR). These laboratory codes become an important part of the nomenclature of laboratory mice and rats because they serve both as identification and as a history of where the animals have been. More information may be found at: http://dels.nas.edu/global/ilar/Lab-Codes.

Outbreds

Outbred stocks of mice or rats are closed colonies of animals, perpetuated by deliberate breeding to maintain the maximum possible heterozygosity within the colony. Outbred mice and rats are named with capital letters. A laboratory code precedes the capital letters and is separated from them by a colon. There are no spaces between any part of the nomenclature. Trademark symbols are not part of official nomenclature.

Examples of outbred stock names:
Crl:CD1(ICR), Crl:CD(SD).

Inbreds

Inbred strains of mice or rats are animals that can be traced to a single ancestral pair and have been produced by at least 20 generations of sibling (or younger parent to offspring) matings. Most inbred strains currently in use have been maintained via brother x sister matings for much longer. These animals are as genetically alike as possible, given normal rates of mutation and genetic drift. Generations of inbred animals are tracked by "F" (or filial generation) numbers, such as F234. Although these are not part of the official nomenclature, records of the number of generations animals have been inbred should be maintained as part of normal colony record keeping. Inbred strains of mice and rats are named with a short string of capital, Roman letters. Names should start with a letter, but the inclusion of numbers is acceptable. For every rule; there is an exception, and some mouse and rat names do not meet these criteria because their names were established before the current criteria (e.g., BALB/c, 129P2). Care should be taken when naming rats and mice that names do not overlap, although there are historical examples of this occurring.

> Examples of inbred strain names:
> C57BL, BN, NZW, SS.

Substrains

Branches of an inbred strain that have genetic differences (either known or probable) are called substrains. Substrain formation occurs under three conditions. When branches of a strain are separated after the 20th and before the 40th generation of inbreeding, substrains are formed, since some residual heterozygosity is still present in animals at the 20th generation. If animals are separated for 20 generations from a common ancestor, the normal rates of mutation and genetic drift will result in substrain formation. When previously unknown genetic differences are discovered within a strain, this may also result in substrain formation. Substrains are identified by a forward slash following the name of parent strain and then either a number or a lab code. For substrains in current use, not all parent strains are still extant. Currently, if more than one substrain is created by a laboratory, a number should be appended after the strain name and before the lab code, although this convention was not always followed. For example, C57BL/6 and C57BL/10 are both substrains of the inbred strain C57BL. C3H/HeJ is a different substrain than C3H/HeN. It should be noted that this nomenclature convention also has its historical exceptions. For example, DBA/1 and DBA/2 are separate strains and BALB/c is not a substrain (and the "c" is lowercase).

Substrain formation is theoretically infinite, because each time a mouse moves to a different laboratory, a new substrain might be formed, purposefully or inadvertently. Laboratory codes should be accumulated at the end of the substrain designation with each move to a different laboratory. Below is a hypothetical example using the authors and imaginary strains and lab codes. These four substrains are different rats and care should be taken not to confuse them, as easy as that might be.

ZZZ/9Stc

The parental strain of this rat is ZZZ. It is the 9th substrain of the ZZZ rat strain inbred by Scout Chou

ZZZ/9StcBje

She then shared them with Bruce Elder

ZZZ/9StcBjeKrpc

He then shared these rats with Kathleen Pritchett-Corning

ZZZ/9StcBjeKrpcLac

Who next shared them with Laura Conour

F1 and F2 hybrids

Another commonly seen mouse or rat is an F1 hybrid. F1 hybrids are made by mating two inbred strains. F1 mice are composed of exactly 50% of each parental genome. They can receive tissue transplants from either of the parental strains and are identical to each other, as long as the same cross is used. However, the resultant offspring are not self-perpetuating. Crosses of two inbred strains can be formed with either parent being of either strain, and the nomenclature reflects this. In F1 hybrids, the female parent is listed before the male parent. Complete nomenclature for F1 hybrids uses full strain names, for example: (C3H/HeN x BALB/cAnN) F1. Typically, however, they are named using the standard abbreviations of inbred strains, followed by F1. The lab code is appended after F1 and may also need to be included in the abbreviation. For example, these are different mice: B6D2F1 and D2B6F1. In the first case, the offspring named had a C57BL/6 dam and in the second, they had a DBA/2 dam. These are also different mice: B6ND2F1 and B6JD2F1. The first mouse had a C57BL/6N dam and the second, a C57BL/6J dam.

F2 hybrids are made by intercrossing two F1 hybrids. With this cross, the allelic contribution from each parent begins to vary since random assortment of alleles comes into play. They are not self-perpetuating and cannot necessarily take tissue transplants from either contributing strain. Their naming continues the F1 theme. Contributing components are followed by F2. Example: B6D2F2 is the offspring of two (C57BL/6 x DBA/2) F1 animals.

As with all other aspects of mouse nomenclature, the abbreviations of inbred strains are standardized. *(Table 11)* Where necessary, append substrain information after the standard abbreviation. For example, C57 is not the abbreviation for C57BL/6 mice and B6J mice are different from B6Crl mice, which differ from B6Ei mice.

Gene nomenclature

When we dissect the nomenclature of genetically modified mice and rats, the first part of the nomenclature is always the name of the animals. Following that is the name of any genes, whether the genes were spontaneously or deliberately modified. As noted in the basic genetics section, mouse and rat genes are identified in a similar fashion, which is different than human gene nomenclature. Formal gene names differ from gene symbols and we will not discuss formal gene

Table 11: Standard abbreviations of mouse strains

Strain	Abbreviation
129	129 strains
A	A strains
AK	AKR strains
B	C57BL
B6	C57BL/6 strains
B10	C57BL/10 strains
BR	C57BR/CD
C	BALB/c strains
C3	C3H strains
CB	CBA
D1	DBA/1 strains
D2	DBA/2 strains
HR	HRS/J
L	C57L/J
R3	RIIIS/J
J	SJL
SW	SWR

(from http://www.informatics.jax.org/mgihome/nomen/strains.shtml)

names further. Mouse and rat genes are given short (3-10) symbols containing Roman letters and Arabic numerals for ease of discussion. These symbols should not begin with a number, and the letter that begins the symbol should be capitalized. When gene symbols are printed, they should be in italics.

> Genes name: *Kit, Kitl, Tyr, Dock7*

Alleles, or different variants of genes, are identified as superscripts to gene symbols. If the allele superscript is lowercase, the allele is recessive. If the allele superscript is capitalized, the allele is dominant, semi-dominant, or co-dominant. In general, gene names and symbols should stay the same over time; although as genes are discovered to belong to different gene families, they may gain new names. If a gene is first identified based on phenotype, it is usually given a name based on that observed phenotype. Once the gene is cloned, the old functional description is appended to the gene symbol as a superscript, with appropriate capitalization. Genes can be discovered in many ways, including identification of DNA sequence, protein product, or production of a phenotype.

Examples of mouse gene and symbol names changing through time (there may have been other, intermediate names in between the two chosen):

c = albino
Tyr^{c} = tyrosinase (note the lowercase "c")

m = misty
Dock7^{m} = dedicator of cytokinesis 7

Dws = dominant white spotting
Kit^{W} = kit oncogene (note the uppercase "W")

Transgenics

Mice with a stable, experimentally introduced foreign DNA sequence are known as transgenic mice. They are named as follows: first is the name of the strain into which the foreign DNA was inserted, followed by a dash. After the dash comes information on the gene inserted into the animal; in this case, the gene name is preceded by "Tg." The inserted gene follows in parentheses. The inserted gene should be named using the official gene symbol for the gene in the species of origin. Promoter designations are encouraged in transgenic lines that differ by tissue expression. Following the gene is a laboratory assigned number, which is often just the nth founder (germline transmission) produced by the lab. Everything after the dash should be in italics and there should be no spaces between the parts. Substrain rules also apply here, so substrain designations may be seen more than once. Only transgenic lines that are maintained need formal nomenclature, and nomenclature may be abbreviated in publications after its initial use.

Example:
C3H/HeN–Tg(DISC1)43Krpc

Explanation:

Mouse	C3H/HeN
Kind of genetic modification	Tg
Gene	*DISC1* (human gene)
Lab identifier	43
Lab code	Krpc

Targeted mutations

Mice with targeted mutations in their genomes are also known as knockouts. These mice are produced by injection of genetically modified embryonic stem cells (ES cells) into gestational day 3.5 embryos. If the ES cells are a different background than the embryo, the contribution of two genomes is noted in the nomenclature. The host (embryo) strain is listed first by its standard abbreviation. The host strain is separated from the ES cell strain with a semicolon or period. Animals will have a semicolon at generations 1-4 of backcrossing to the embryo donor strain, and a period at 5 or greater generations of backcrossing to the embryo donor strain (incipient congenic, or congenic). The targeted gene symbol and the rest of the information on the targeted mutation is separated from the mouse strain information with a dash and italicized. The remaining identification is superscripted to the gene name (because it is an allele of the gene): "tm" for targeted mutation, a numeric designation given by the laboratory, and the laboratory code.

Example:
C3N;129P2–*Disc1*^{tm83Krpc}

Explanation:

Embryo	C3H/HeN
ES cell background	129P2/OlaHsd
Gene	*Disc1* (mouse gene)
Kind of genetic modification	*tm*
Lab identifier	*83*
Lab code	*Krpc*

Congenics

Congenics are animals produced by repeated backcrosses to an inbred strain. They differ from the host strain by one or more genes donated from another strain(s). Congenics can be useful to study how genes and alleles behave when transferred from one background to another. In congenic nomenclature, the background strain is listed first and usually abbreviated. The donor strain is separated from the background strain by a semicolon or period. Animals will have a semicolon between the two background abbreviations at backcross generations 1-4 and will have a period at 5 or greater generations of backcrossing (incipient congenic). Mouse strain information is separated from the transferred gene information by a dash. The transferred gene is in italics and, remember, it can be any gene. A substrain designation may follow after a forward slash. Backcross generation numbers are not part of the formal nomenclature, but should be included to assess the potential for heterozygosity. Congenic animals will have two different generation numbers, an "N," indicating the number of backcrosses, and an "F," indicating the number of intercrosses after the backcrossing was finished. For example, an animal may be N5 + F32, meaning that it has been intercrossed for 32 generations after the 5th backcross.

Congenic:

Example:
C3N;D2–*H2*^{d}

Explanation:

Background	C3H/HeN
Punctuation	semicolon (N1-4)
Donor	DBA/2
Gene	*H2*^{d}

Spontaneous mutation congenic:

Example:
C3N.129P2–*Disc1*^{rcm}

Explanation:

Background	C3H/HeN
Punctuation	period (N≥5)
Donor	129P2/OlaHsd
Gene	*Disc1*^{rcm}

Transgenic congenic:

Example:
B6Crl;D2–Tg(DISC1)43Krpc

Explanation:

Background	C57BL/6NCrl
Punctuation	semicolon (N1-4)
Donor (original background)	DBA/2
Kind of genetic modification	Tg
Gene	*DISC1* (human gene)
Lab identifier	43
Lab code	Krpc

Targeted mutation congenic:

Example:
C3N.129P2–*Disc1*^{tm83Krpc}

Explanation:

Embryo	C3H/HeN
Punctuation	period (N≥5)
ES cell background	129P2/OlaHsd
Gene	*Disc1* (mouse gene)
Kind of genetic modification	*tm*
Lab identifier	*83*
Lab code	*Krpc*

More information to clarify mouse or rat background strains may be included in nomenclature. These include the notations "Cg" and a strain abbreviation in parentheses [e.g., (B6)]. When Cg is noted in the nomenclature, the gene or genes being introgressed are on a mixed or complicated background. For example, a rat congenic for a mutation arising in an outbred stock would have Cg as its second background symbol. When parentheses are used, there is a known contribution from a third strain.

Examples: BN.Cg-*Lol*^{wut}*, C3H.C(B6)-Disc1*^{tm83Krpc}

There are three other types of genetically altered mice that are relatively often seen in laboratories. These include coisogenic, consomic, and conplastic animals. A coisogenic mouse or rat is formed by the occurrence of a mutation at a single locus within an inbred strain. Coisogenic animals are named with the strain symbol (and substrain symbol where appropriate), followed by a hyphen and the gene symbol of the mutated allele in italics.

Example: C57BL/6NCrl–*Lol*^{bbq}

Consomic animals are also known as chromosome substitution strains. They are created by repeated backcrossing of a whole chromosome onto an inbred strain. As with congenics, a minimum of 10 backcross generation is required. The generic designation is HOST STRAIN–Chr $\#^{\text{DONOR STRAIN}}$. There is a space between Chr and the chromosome number.

Examples: C57BL/6J–Chr $13^{\text{DBA/2J}}$, SS–Chr 4^{BN}

In conplastic animals, the nuclear genome (i.e., mitochondrial genome) from one strain has been crossed onto the cytoplasm of another. Their generic designation is NUCLEAR GENOME–$\text{mt}^{\text{CYTOPLASMIC GENOME}}$.

Example: C57BL/6N–$\text{mt}^{\text{BALB/c}}$

PART 6 Production and maintenance of colonies

Production planning

Colony production and management starts with production planning, thus it is important to understand both the purpose and the goals of the breeding colonies. For example, is the purpose to supply females for embryo or blastocyst harvest, to supply recipient females for embryo transfer, to generate novel animal lines for further characterization, to supply small or large scale production for specific experiments, or to serve as a back-up colony as part of disaster planning? In addition, what are the goals for the breeding colony in terms of number of animals needed over a specific period of time, and the characteristics of the animals needed, such as sex, age range, and genotype? Finally, other factors that can influence production performance should be considered during the planning process. Factors, such as the background strain characteristics (e.g., fertility rate, litter size, litter frequency), maternal characteristics, breeding systems and mating scheme, number of available breeders, model phenotype and breeding life span, and general health status of the colony will affect a breeding colony's overall productivity.

Breeding system and mating scheme

The breeding system, or breeding cage, may be placed in one of two categories. Advantages and disadvantages associated with each system have been described.[175] One category is permanent mating groups, to include monogamous and harem (one male with more than one female) matings. In permanent mating groups, the male is housed continuously with the female(s) which allows him to participate in pup care and to take advantage of the postpartum estrus. Alternatively, there are temporarily mated groups, which include polygamous (multiple males and females) and observed mating (a.k.a. hand mating or timed-mating). These require separation of the breeders at some point after mating.

To maximize the productivity of female mice, they are best kept in permanently mated groups because this allows mating at the postpartum estrus. Although male intensive, the monogamous breeding system will result in the most number of pups born per female over her reproductive lifespan. On the other hand, harem breeding will result in the most number of pups born per breeder cage, but at the expense of decreased individual female output.[176] While multiple breeder females housed in the same cage may share pup rearing tasks, it will also make record keeping more challenging without vigilant monitoring of the breeding units. Polygamous mating is the least male-intensive breeding system, but because pregnant females are separated from the males to litter in separate cages, it leads to the fewest number of pups born per female, and record keeping can be difficult because male parentage is not certain. To maximize the productivity of a single male mouse, it may be best to rotate different receptive females into his cage. This can allow for accurate staging of gestation and is commonly used for generating time-mated females for embryo harvest, or with vasectomized males ("duds"), to prime the uterus of females for embryo transfer surgeries.

There are many possible mating schemes for breeding genetically modified rodents, but not all of them can be used for the maintenance and propagation of animals carrying a particular genetic modification. Any mating scheme should take into consideration the genotypes of the breeders to generate offspring with the desired genotype for subsequent use. This is especially important if a phenotype of interest is expressed only in homozygotes or if expression is sex-dependent. Therefore, having a strong foundation in genetics can be helpful when managing rodent breeding colonies.

Breeding two homozygotes will yield 100% homozygous offspring and is useful if the gene effect is seen only in homozygotes and if homozygotes are viable and fertile. Although there will be no sibling control animals, inbred animals of the same strain may be used if the mutants are on an inbred background. Mating of a homozygote with a heterozygote will yield 50% homozygotes, while the other 50% will be heterozygote siblings. This scheme is useful when the phenotype is seen only in homozygotes, and when littermate controls are required. This mating scheme may also be chosen when one sex of homozygotes is not viable or fertile. Mating two heterozygotes will produce 25% homozygote, 50% heterozygote, and 25% wild-type offspring. This scheme is useful when homozygotes show the desired phenotype but are not fertile. This mating scheme is also useful if heterozygotes are of interest because they have a phenotype intermediate between wild-types and homozygotes. Mating of a wild-type and a heterozygote yields 50% wild-type and 50% heterozygous offspring. This mating scheme is useful with animals that have sex-linked mutations. When expected percentages of genotypes are given, it is probable that genotypes will appear in this ratio in offspring produced over time. It is not a guarantee that every litter will have a particular combination.

Other breeding practices to be aware of, especially when managing mutant colonies or transferring a mutation onto inbred strains, are backcross, incross, intercross, and outcross, and cross-intercross matings. A backcross is a cross between one animal type that is heterozygous for alleles obtained from two parent strains and a second animal type from one of those parental strains. An example would be the crossing of a mutation into an inbred background, a common practice when generating congenic strains. An incross is a cross between two animals with the same homozygous genotype at the designated loci. Standard lines of inbred animals are maintained via incrossing. An intercross is a cross between two animals with the same heterozygous genotype at designated loci. An outcross is a cross between genetically unrelated animals such as crossing a female C57BL/6 and a male DBA/2 to generate B6D2F1 hybrids.

Here is an example of transferring of a homozygous mutant allele from a BALB/c (C) background to a C57BL/6 (B6) background, followed by maintenance of the mice as homozygotes:

OUTCROSS:	BALB/c-Abc^{pdq} x C56BL/6 yields CB6F1 heterozygous for Abc^{pdq}
BACKCROSS:	Abc^{pdq} heterozygotes are mated back to wild-type B6 for 10 generations N2 – N4: B6;C–Abc^{pdq} N5 – N10: B6.C–Abc^{pdq} (incipient congenic)
INTERCROSS:	B6.C–Abc^{pdq} heterozygotes at N10 are mated with each other to generate homozygotes
INCROSS:	B6.C–Abc^{pdq} homozygotes are mated with each other

Meeting production expectations

It is important to be able to justify animal needs to an institution's animal care and use committee by logically demonstrating how a desired animal number was reached. When calculating animal numbers, consider all the key functional groups within a breeding colony, including donor/recipient females, stud/dud males, or breeders/future breeders, stock (experiment) animals, whether animals are commercially available or need to be produced in-house, and how many of each population will be needed.

Historic information regarding the reproductive characteristics of different basic strains and stocks are readily available.[177,178] These data should be considered starting points when working with animals of various backgrounds. Normal data also change through time due to changes in health status, genetic drift, and environmental factors, as evidenced by *Table 12*. Ideally, data collected from colonies of interest at institutions will allow for accurate calculation of the reproductive performance of particular stocks or strains. Such data can also be used to help manage production expectations, plan studies, and troubleshoot production problems when needed. Basic productivity information should include pregnancy rate (percent females pregnant per week), average litter size, interval between litters, number of cages retired due to infertility, pup survivability rate (pups weaned of total pups born expressed as percentage), observations on the females' ability to rear pups (e.g., note incidence of cannibalism, milk production, neglect of litters, etc.), and special housing/health considerations.[179]

Table 12: Litter size of C57BL/6J mice through time. Data collected from editions of The Jackson Laboratory Handbook on Genetically Standardized Mice.[178]

Year reported (years collected)	Litter size born as reported by The Jackson Laboratory
2009 (05-07)	5.9
1997 (89-90)	6.6
1991 (86-87)	6.8
1982 (80-81)	7.0
1980 (78-79)	6.7
1968 (63-65)	5.9*
1962 (60-61)	6.1*

*Information listed as "litter size". Detail not provided as to whether this is born or weaned litter size.

Production Index (P.I.) is calculated by dividing the number of weaned animals of both sexes by the number of females bred in the colony during a fixed time period.[179] Mathematically this may be written as "# pups weaned/#females/week." The following example demonstrates how to calculate the P.I. for a colony:

Scenario: A breeding colony comprised of 20 trio (1 male: 2 female) breeding cages weans, on average, about 60 pups of both sexes each month.
P.I. = # pups weaned/#female/week → P.I. = 60/40/4 = 0.375

If there is no historical production data to work from, for general estimation purposes, assign a P.I. of 2 to outbred stocks of rats and mice, 0.8 to "good" inbred mouse breeders and most inbred rats (i.e., FVB, LEW), a P.I. of 0.5 to "average" inbred mice breeders (i.e., C57BL/6), and a P.I. of 0.3 to "poor" breeders (i.e., DBA, BN). More information on mouse and rat strains and stocks may be found in *Table 13.*

Table 13: Production indices for various Charles River stocks and strains of rats and mice

Stock/Strain	Production index
C57BL/6	0.5
C3H	0.8
BALB/c	0.8
DBA/2	0.35
FVB	1.0
Swiss origin (outbred)	2.0
Outbred nudes	1.0*
Inbred nudes	0.5*
CB17-scid	0.8
CD	2.0
WH	2.0
F344	0.8
BN	0.4
LE	2.0

*Production index in nudes is based on nude animals produced so is half of true production index

The production index is used to estimate the number of breeding females needed to supply specific cohorts of animals for experiments. To sustain a continuous supply of animals (i.e., weekly or every other week) or if the cohort age range is narrow, the basic mathematical formula to use may be written as:

# of breeding females x P.I. = # of pups weaned/week

When performing colony size calculations, take into consideration the characteristics of the animals needed for experiments, such as age, sex, and genotype, as well as productive mating frequency (or fertility rate), and the need for replacement breeders. The colony must be self-sustaining for continued production. In addition, the calculations below do not take into account the bolus of offspring when matings are all set up at once. Using information from Tables 1 through 3, here are additional colony production calculation examples to demonstrate how to use this basic formula:

Scenario

An investigator asks for 50 6-week old female KO mice for a study. The source colony is homozygous knockout mice on an FVB/N background. How many breeder females are needed?

Productive mating frequency for FVB/N = 90%

Average litter size = 9.5 pups

Colony is being managed as HO x HO

Assume equal offspring sex ratio and minimal pup mortality
If 50 females are needed, then it is necessary to produce at least 100 total pups/week.

At 9.5 pups/litter, need to produce at least 11 litters to meet demand.

With 90% productive mating frequency, 13 breeder females should be set up at one time.

Scenario

Now the investigator wants a continuous supply of 20 males, 6-7 weeks old, every other week, for a series of planned experiments. How many active breeder females are needed to maintain this production?

If 20 males are needed every other week, then 10 males need to be produced every week.

If 10 males/week are needed, then it is necessary to produce at least 20 total pups/week.

# of breeding females x P.I. = # of pups weaned/week → # females x 0.8 = 20 pups/week

females = 20/0.8 = 25 productive females needed.

With 90% productive mating frequency, need to maintain at least 28 breeder females in the colony. (25/0.9=27.8)

Scenario

When should this investigator expect her first batch of study animals if there are only 10 breed age females and 2 males available in the colony?

Week #1: rotate breed age females through the stud cages and remove them post mating

Week #4-7: following 3 weeks gestation, 86 pups are born. (10 females x 0.9 x 9.5 pups/litter); 50:50 sex ratio, 43 females and 43 males are weaned and set aside as breeders

Week #12: Set up active breeding colony using 8-week-old sibling animals

Week #15: First batch of study animals born

Week #18: First batch of study animals weaned

Week #21: First batch of 6-week-old study males delivered to investigator for study

Please note: While the investigator could have used 20 of the homozygote males at week #10, there would not be a sufficient number of age-appropriate animals for subsequent studies during week #12, nor would there be enough males in the colony to sustain production needs.

For examples and recommendations on animal number requirements when breeding mutants for genetic analysis, please refer to *Guidelines for the Care and Use of Mammals in Neuroscience and Behavioral Research* by the National Research Council.[180] Experiments that involve the determination of genetic inheritance pattern, identification of genes involved in a quantitative trait, or fine mapping to determine chromosomal location of mutant gene, can be very animal intensive. For example, approximately 1200 total animals are needed to map a single gene with recessive inheritance pattern and full penetrance. This number includes progeny animals for developmental, phenotyping, and gene linkage analysis studies, and assumes that the breeding colony is comprised of 10 to 12 pair-mating cages, with breeders that exhibit no unusual adverse, life-interfering phenotypes, have good productive mating frequencies, and possess an average reproductive life span of 6 to 8 months. In contrast, approximately 80 to 100 animals may be required to characterize a line from founder animals, operating under the assumption that up to five breeder pairs are maintained per line created and that appropriate numbers of weanlings are available for genotyping and phenotyping.

Breeder selection and replacement

In general, the productivity efficiency will drop as breeding animals age, as demonstrated in *Table 14*. The typical reproductive life span of mice is approximately 6 to 8 months, but breeder males may be used for longer, if necessary. The breeding cycle of mice in commercial production is set at approximately 6 months after set-up, while published information on the optimum breeding period of rats is limited (about 9 months is usual).[176] When managing your colony, it is recommended that the average breeding and replacement cycle should be set based on the colony's production history.

Table 14. Production efficiency index of nine inbred strains of mice paired for different periods. Table adapted from Festing and Peters, 1999.[176]

Strain	10 weeks	17 weeks	23 weeks
AKR	0.60	0.63	0.55
BALB/c	0.63	0.74	0.79
CBA/Ca	0.98	1.17	1.11
C57BL/Lac	0.41	0.62	0.64
DBA/1	0.85	0.81	0.69
DBA/2	0.58	0.58	0.54
NZW	0.58	0.59	0.69
Average value	0.66	0.73	0.71

Animals that exhibit the standard phenotype reported should be selected as breeders. For example, if maintaining mice on a B6 background, avoid selecting animals that come from large litters, or have parents that are known to have produced offspring with background strain lesions (e.g., hydrocephalus, malocclusion, or microophthalmia). Make sure a breeder rotation schedule is established to sustain the colony size while meeting short- and long-term production goals. This rotation schedule will depend on the effective reproductive lifespan of the animal model. A suggested rotation schedule is provided in *Table 15*. Monitor the reproductive performance of the breeder animals and replace those that are under-performing (i.e., no litters born after 2 to 3 months of set-up, prolonged interlitter intervals, consecutive loss of 3 litters due to inability to raise or wean pups), have obvious health issues, or express unfavorable phenotype(s).

Record keeping and colony organization

The ability to keep organized breeding records is vital to the success of colony management. Breeding cages should be checked at least once per week, and basic breeding information can easily be tracked at cage side. Collected information is often transferred into an electronic database as part of laboratory record maintenance. The database can also be used to calculate and track specific reproductive parameters, facilitate production performance evaluation,

Table 15. Breeder retirement standards. Table adapted from Guidelines for the Care and Use of Mammals in Neuroscience and Behavioral Research, 2003.[180]

Effective reproductive lifespan	% of colony replaced monthly
5 months	20
6 months	16.7
7 months	14.3
8 months	12.5
9 months	11.1
10 months	10

and identify problems or corrective measures during troubleshooting exercises. Failure to stay organized could result in genetic contamination of the colony, decrease in production, difficulty in locating specific animals, or delays in experiments.

Animal identification

Individual animal identification may be achieved through multiple different methods. Individually numbered rodent ear tags and tag applicators are commercially available through companies such as National Band and Tag (www.nationalband.com) and Hasco Tag Co. (www.hascotag.com). An ear punch or ear notch method may be used, and special punch applicators for mice and for rats may be found in scientific or surgical instrument catalogs such as Kent Scientific (www.kentscientific.com) and Roboz (www.roboz.com). Example ear clipping systems are illustrated in *Figure 28*. Toe clipping is a controversial identification method that requires scientific justification, and it is banned in some institutions. Current literature on toe-clipping describes limited effects in mice if certain conditions are followed.[181,182] It is easy to mark neonates by this method, but it should be done with clean, sharp scissors on pups approximately 7 days of age, only the distal phalanx should be cut, and no more than one toe per foot should be amputated. Single use, implantable microchips are commercially available through companies such as Biomedic Data Systems (www.bmds.com) or Topaz Technologies (www.topazti.com). Finally, animals may be

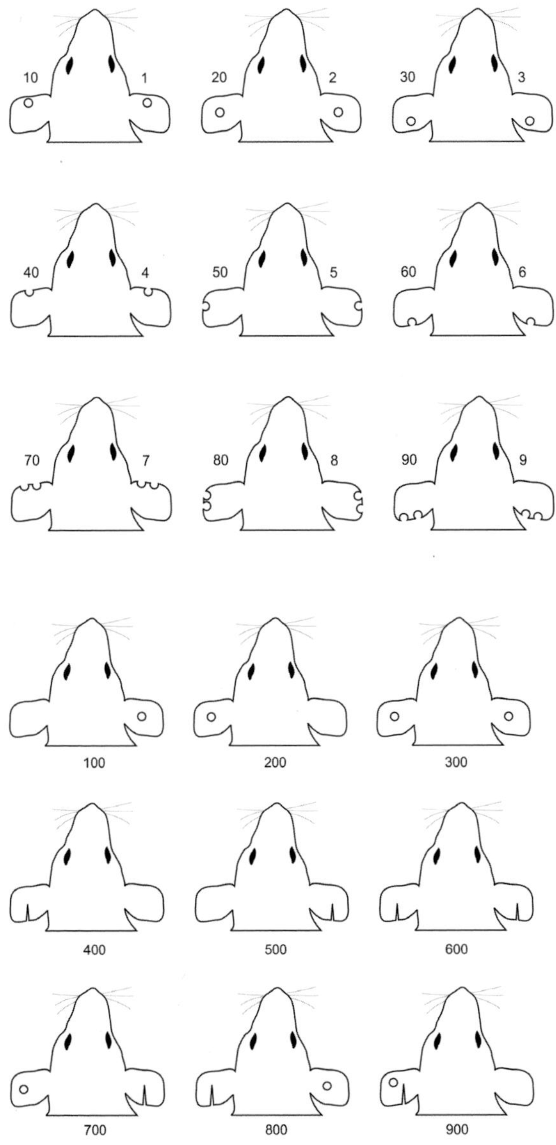

Figure 28. Examples of ear clipping or punching systems for numbering mice and rats. Redrawn from Dickie, MM. in Biology of the Laboratory Mouse (ed E. L. Green) p23-27 (Dover, 1975) and Inglis, J.K. Introduction to Laboratory Animal Science and Biology p51 (Pergamon Press, 1980).

tattooed on the tail, paw or individual digits using products sold by companies such as Ketchum (www.ketchum.ca), and Animal ID and Marking Systems (http://www.animalid.com/), or Somark Labstamp (www.somarkinnovations.com). See *Figures 29 and 30* for examples of tattoo marking systems. These identification methods may be used alone, or in combination. *Figure 31* shows a combined ear punch and tattoo system. Whichever system is used, remember to be consistent, stay organized, offer proper training to the animal care staff, and include information on the method when sharing animals with other institutions or colleagues. Benefits and drawbacks of each method are presented as *Table 16.*

Cage level identification

Specialty cage cards should be used based on the purpose of the animals in the cages. For example, stock animals, breeders, embryo donors/recipients, dud/stud, etc., and proper nomenclature should be used. On a breeder unit's cage card, relevant information may include: identification and date of birth of the breeders, date of breeding unit set-up, age of the breeders at first litter, the dates of all litters born/weaned, and the expected retirement age/date for the breeding unit. Once a breeding cage is retired, the retired breeder cage cards should also be filed away as part of the colony record.

Laboratory records

It is often best to have multiple places to find and cross-reference colony information. Complete colony records should also include genotyping data. Many electronic software and programs are available that can be used to manage rodent colony databases. A simple lab notebook will also work. Some examples, but by no means an exhaustive list, include Microsoft Excel, Microsoft Access, Colony by Locus Technology (www.locustechnology.com), Big Bench Mouse (www.bigbenchsoftware.com), Scion by TopazTracks (www.topazti.com), Progeny (www.progenygenetics.com), JCMS (colonymanagement.jax.org/), MausDB (www.jupiter.helmholtz-muenchen.de), PyRAT (www.scionics.de/pyrat).

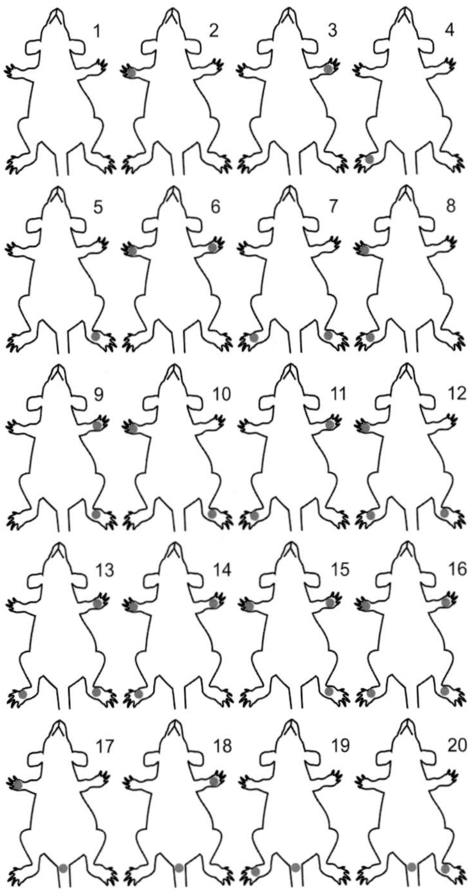

Figure 29. Examples of toe and foot tattooing systems for identification of mice and rats. Redrawn from Hetherington, C. M., Doe, B. & Hay, D. in Mouse Genetics and Transgenics: A practical approach The Practical Approach Series eds Ian J Jackson & Catherine M Abbott) p1-26 (Oxford University Press, 2000).

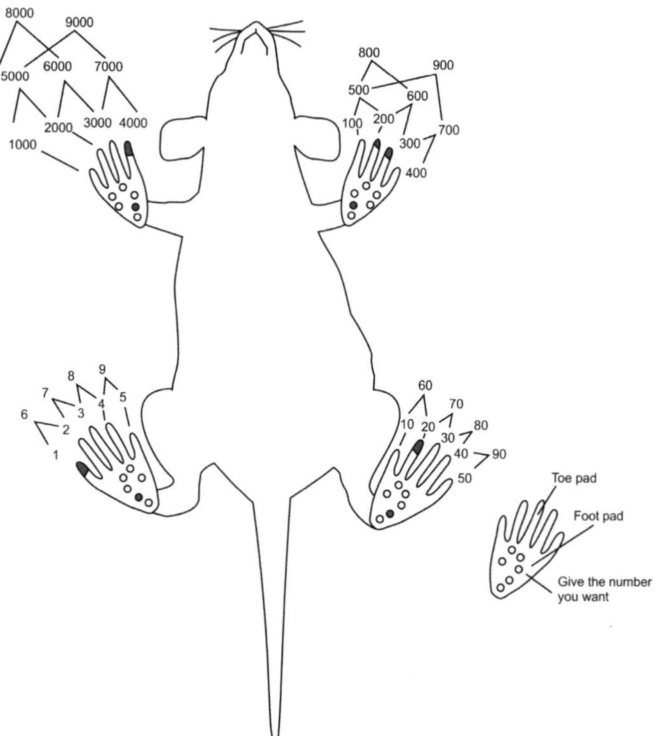

Figure 30. More examples of toe and foot tattooing systems for identification of mice and rats. Redrawn from Hetherington, C. M., Doe, B. & Hay, D. in Mouse Genetics and Transgenics: A practical approach The Practical Approach Series eds Ian J Jackson & Catherine M Abbott) p1-26 (Oxford University Press, 2000).

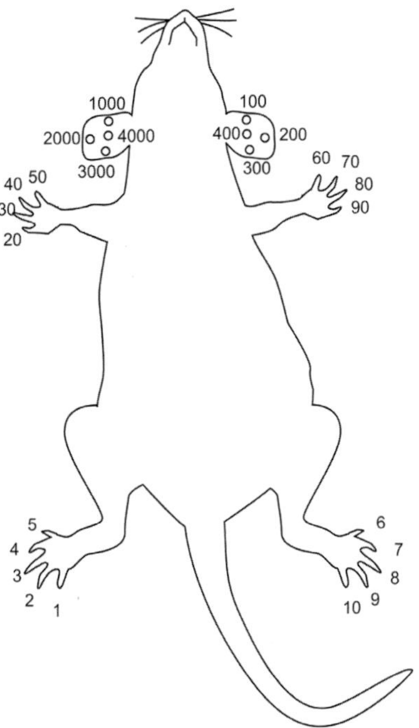

Figure 31. Example of a combination ear marking and toe tattooing system for mice and rats. Redrawn from Hetherington, C. M., Doe, B. & Hay, D. in Mouse Genetics and Transgenics: A practical approach The Practical Approach Series eds Ian J Jackson & Catherine M Abbott) p1-26 (Oxford University Press, 2000).

Table 16. Commonly used methods of identification and reasons for and against their use

Method of identification	Pros
Ear tag	Inexpensive Easy to apply User can specify number/letter combinations Machine-generated numbers and letters are usually clear Variety of materials, including optical reader compatible tags
Ear punch or notch	Inexpensive Quick to apply Sample for genotyping generated at the same time as identification Many variations on the system available for identification
Toe clip	Inexpensive Allows for identification of neonates Sample for genotyping generated at the same time as identification If performed within certain parameters, recent literature indicates minimal sequelae
Microchip (RFID transponders)	Difficult to misinterpret identification of animals Microchip can be retained with samples Some microchips can be reused Animals cannot obscure microchip
Tattoo	Permanent Very clear identification of animals Animals cannot remove tattoo Can be performed in neonates No special equipment required

Cons
Can fall out May become infected[183] Can potentiate tumor formation[184,185] Cannot be applied before ear is fully developed Separate genotyping sample must be taken May be painful
Cannot be applied before ear is fully developed Can be rendered useless by fighting or normal healing[186] May be difficult to interpret May be painful Requires skill to perform in conscious mice
Banned by many institutions Should not be applied to animals younger than 3 or older than 7 days[182] May be painful Motor defects are possible after phalanx removal[182] Valid ethical limitations on number of toes removed per paw limits numbering system The toe may grow back[187]
Expensive when compared to other methods Requires reader Number determined by manufacturer Difficult to apply to neonates Requires anesthesia Separate genotype sample must be taken Can induce tumor formation[188,189]
Equipment helpful for some types of tattoos Can be difficult to read in pigmented animals Separate genotype sample must be taken May be painful Tattoos in neonates can fade[190] Tattoo pigment can be found in regional lymph nodes[191]

PART 7

Clinical assessment

Performing regular clinical observations on colony animals is crucial to effective colony management, especially when managing genetically altered animals and characterizing novel animal models. Since genetic manipulations can sometimes lead to unpredictable outcomes (i.e., unexpected phenotype), documentations of specific clinical signs followed by proper diagnostic workup on principal and sentinel animals, pathologic assessments, and elective phenotyping assays should be performed to determine whether specific clinical signs observed are related to animal health status, their genetic background, or phenotypes resulting from genetic modification.

All animal users should receive training on how to recognize normal and abnormal rodent behavior and appearance, and all are encouraged to spend time in animal rooms to gain a better appreciation of normal rodent appearance and behavior. In addition to clinical sign recognition training, an institution should have a system in place so abnormal findings are reported. Timely reporting of clinical signs or colony performance-related problems such as an increase in mortality, decrease in frequency of expected genotypes, or decrease in production, should be made to both the veterinary and the investigator groups to initiate diagnostic workups and troubleshooting exercises. As a training tool on how to perform clinical observations and for examples of clinical signs that may occur in a rodent colony, please refer to the *Charles River Handbook on Clinical Observations of Rodents and Rabbits.*[192]

Known phenotypes associated with common strains and stocks of mice and rats are available online. For rats, information may be found at the Medical College of Wisconsin's Rat Genome Database (rgd.mcw.edu/wg/physiology?100) and at the National Bioresource Project for the Rat in Japan (www.anim.med.kyoto-u.ac.jp/nbr/phenome/).The Mouse Phenome Database (phenome.jax.org) is a searchable database curated by the Mouse

Genome Informatics Group at The Jackson Laboratory and contains information submitted by scientists from around the world. Dr. Michael F. W. Festing's work, *Inbred Strains of Mice and Rats and their Characteristics*, can be searched here (www.informatics.jax.org/external/festing/search_form.cgi). Dr. Festing's list was last updated in 1996, but remains an excellent resource. All of these sources, as well as others, can provide rodent colony managers with useful animal model information and can help decipher whether an unusual clinical observation is considered a background lesion, an animal health concern, or a novel phenotype worthy of further characterization. A short list of some normal background lesions and known mutations in some inbred mice is provided in *Table 17*.[193]

Table 17: Background strain characteristics

Strain
C57BL/6
BALB/c
C3H/He
129 strains
FVB
Swiss-origin outbred mice, Swiss Webster, ICR, NIH, CD1, CFW, Black Swiss

Characteristics
Ophthalmologic abnormalities (micro- and anophthalmia, cataracts)[194]
Malocclusion[195]
Hydrocephalus[196]
Ulcerative dermatitis and barbering
Age-related hearing loss (*Cdh23*^{ahl})[72]
Splenic melanosis
Senile amyloidosis[197]
Susceptible to preputial adenitis
Absence of corpus callosum in about 30% of mice[198]
Dystrophic mineralization[199]
Accessory adrenal cortical nodules
Age-related hearing loss (*Cdh23*^{ahl} in the ByJ substrain)[72]
Two main substrains: C3H/HeN and C3H/HeJ
Retinal degeneration caused by *Pde6b*^{rd1}
Dystrophic mineralization[199]
Alopecia areata in aged C3H/HeJ mice[194]
C3H/HeJ has a mutation in *Tlr4*, so doesn't respond to endotoxin[200]
Many substrains; be careful when matching ES cell lines to mouse backgrounds and of historical name changes[201]
Mutation in *Disc1*[202]
Testicular teratomas (common in 129X1 –formerly 129/SvJ; incidence differs between substrains)
Absence of corpus callosum in about 70% of mice of 129P3 substrain (formerly 129/J)[198]
Age-related hearing loss (*Cdh23*^{ahl} in 129P1/ReJ and 129X1/SvJ)[72]
Retinal degeneration due to *Pde6b*^{rd1}
Prone to seizures with neuronal necrosis
High tumor incidence in aged mice (pituitary adenomas, alveolar-bronchiolar tumors, hepatocellular tumors, Harderian gland adenomas, etc.)
Different stocks have different frequencies of *Pde6b*^{rd1} allele[62,63]
CFW have a high prevalence of lymphoma due to endogenous retrovirus[203]

PART 8 Troubleshooting colony performance

When trying to determine why a breeding colony is not performing well, stating the problem can often be the most difficult step. What is truly the problem with the colony? Not having enough animals for experiments may not always be the fault of the animals. Disorganization, inappropriate mating schemes, health problems, unanticipated phenotypes, and environmental challenges can all affect colony performance. There are many things that can go wrong during the long, complicated processes of breeding, pregnancy, parturition, and development. In general, however, domestic mice and rats in the laboratory have been selected for their ability and willingness to breed in captivity. They breed in spite of what we do, not because of it.

> What's the real problem?
> "My mice don't breed."
> "My mice don't get pregnant."
> "My mice get pregnant, but I don't see any pups."
> "My mice deliver pups but the pups die."
> "My mice breed, get pregnant, and raise their pups, but the pups die before/at weaning."
> "I'm not getting enough pups."
> "These mice don't look right."

Where to start

The first step in defining, investigating, and solving the problem is to review the breeding records. Breeding and production records must be organized and complete, with sufficient details recorded so that analysis of these data can be performed.

The following indices should be recorded:

Necessary for normal organization:
- Disposition of animals (breeding, death, experiment)
- Date of birth of breeders and stock
- Date of set-up for breeders
- Generation or backcross number
- Proper nomenclature
- Breeding chart or pedigree book or program

Crucial to know how the colony is performing:
- Average litter size by litter number
- Time interval between litters
- Number of cages retired due to failure to produce
- Pup survivability rate (% pups weaned/total pups born)

Nice to have:
- Time to first litter after set-up
- Observations on the females' ability to rear pups

Knowing the final disposition of an animal is important so no effort is wasted determining whether an animal is really alive or only alive in the computer database. Recording deaths of breeders will permit you to compare the lifespan of the animals in your breeding colony to life spans published in the literature or available from the vendor. Additional dispositions for mice or rats within the colony should include:

- Assigned to study
- Found dead
- Euthanized due to clinical issue
- Euthanized – retired breeder

Good, consistent record keeping will minimize lost animals.

When troubleshooting production problems, take time to go into the vivarium and look at the animal and its environment. Often, the cause becomes evident in a brief examination. For example, are the animals breeding "on paper" really set up in the cage? Removing animals from their cages and confirming identification will clear up record keeping problems. Ensuring that animals of opposite sexes are set up to breed can explain

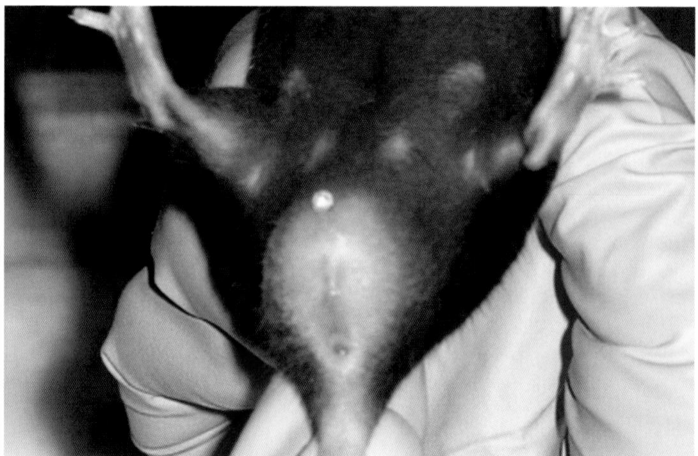

Figure 32. A female mouse with an imperforate vagina. In this mouse, the vaginal closure membrane failed to open at puberty. The normal uterine and vaginal secretions accumulated behind the membrane, resulting in distention of the vagina. This animal may appear male at first glance, but male mice and rats do not have nipples.

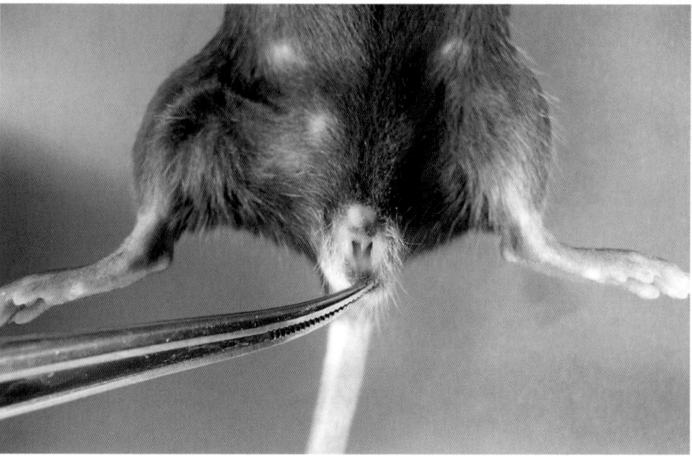

Figure 33. A female mouse with a vaginal septum. This mouse is unsuitable for breeding because the structure present in the vagina will interfere with both mating and parturition.

an unproductive cage. *(Figure 32 and 33)* Before setting any pair of animals up for breeding, a quick physical exam can rule out problems such as vaginal septum or imperforate vagina. Other items easily checked include breeders past their prime, missing breeders, and lack of young animals set aside for breeding. Knowledge of reproductive lifespan and expected production based on background of the animal is important to managing production expectations. A BN rat will never have a litter size of 12, and a DBA/2 mouse will never have a litter size of eight. Besides background, awareness of the effect of phenotype on reproductive lifespan is critical to managing breeding efficiency in a colony.

Easy fixes for production problems:

- Male and female(s) mating?
- Healthy enough to mate and physical conformation allows for mating?
- Animals too old?
- Correct identification of breeding animals?
- Correct animals breeding?

Comparison of genotyping results across multiple litters should approach Mendelian frequencies. If one litter has no homozygotes, it is probably due to chance. If five litters have no homozygotes, another explanation should be sought. Although embryonic lethality in homozygotes is a possibility, the more prosaic and likely explanation is a mistake in breeder selection. Causes of this could be the selection of animals with the wrong genotype or incorrect genotyping results. If no homozygotes are being produced, and the Mendelian distribution of pups is correct for a heterozyote x heterozygote mating, this will more likely be a developmental problem with pups. If the allele frequencies are typical of a heterozygote by wild-type mating, check the genotypes and identification of breeders.

Strain effects

Strain effects influence breeding and production. Strain effects are those features of particular inbred strains that affect their reproductive performance, whether through biology or behavior. For example, male SJL/J mice are notorious for aggressive behavior. Unlike other aggressive strains, like FVB/N, which tend to be male-aggressive, SJL are indiscriminate in their attacks, killing and wounding mates and pups. Phenotypic differences based solely on background can have an impact on behavior, physiology, and breeding.

Genetics

Two other phenomena that can have profound effects on production are inbreeding suppression and hybrid vigor. When inbreeding animals or backcrossing an allele (mutant or otherwise) onto a pure inbred background, a notable decrease in litter size and average pup weight and size will occur from one generation to the next. This phenomenon is most prevalent between F2 and F8 or N2 and N8 and is related to the cumulative burden of fixing deleterious alleles.[204] To continue breeding through this normal event, it is important to maintain breeders from the previous generation of breeding until the reproductive success of the current generation is evident. Backcross breeding paradigms should also be started with a sufficient number of breeders so that infertile pairs can be discarded.

Backcrossing tips:

- Backcross females first to a male of the new strain to fix the Y chromosome
- After the first generation, only select heterozygous males to backcross
- F1 mice will be larger and more vigorous than parental strains
- N2 - N10 mice will begin to take on the characteristics of the new strain
- Keep breeders from the previous generation until you're sure that the next generation will breed and have the genotype
- Consider intercrossing at N5 to see if homozygotes retain phenotype

The phenomenon of hybrid vigor is an increase in fertility and fecundity due to increased genetic diversity. This may also be accompanied by an increase in body size and overall fitness. An increase in litter size and production of robust pups are the defining characteristics of this event. Hybrid vigor is often the first indication of genetic contamination, potentially resulting from inadvertent breeding of multiple transgenic lines or animals from different backgrounds. Changes in vigor of animals and litter size will often be noted before coat color changes or shifts in expected Mendelian frequencies. If a colony of BALB/c mice begins to have an average litter size of 10, there is likely something wrong. Every facility should regularly monitor animals for genetic contamination as well as contamination by infectious organisms. For help in designing health monitoring programs, the *Companion Guide to Rodent Health Surveillance for Research Facilities* can be a valuable resource.[205]

Protecting the genetic investment

- Genetic quality control monitoring using SNP or microsatellite panels
- Cryopreservation of important strains and stocks
- Biological controls
    - Capture and euthanize escapees
    - Keep foundation or breed stocks separate from experimental colonies
- Personnel training
    - Staff working with breeding colonies should understand genetics
    - Staff should know what is normal for that colony and be able to identify deviants
    - Staff should understand the record keeping system
- Administrative controls
    - House strains with different coat colors next to each other
    - Use different colored cage cards
    - Keep good breeding records

Watching the behavior of the animals in their home cage is an underutilized approach to defining the nature of breeding difficulties. Stress-induced stereotypies or behaviors related to neurobehavioral phenotypes may prevent normal copulatory behavior or impair maternal rearing abilities. In the case of an adverse neurobehavioral phenotype influencing pup development and survival, fostering pups is often necessary. When evaluating troubles in production or breeding, it is important to remember that infanticide and cannibalism are normal behaviors in response to stress and dead pups. Efforts should be directed to determine which stressful event may have resulted in pup killing or pup death.

Phenotype

The very phenotype under study can also adversely affect embryonic and neonatal survival as well as fertility. Embryonic lethality is often difficult to differentiate from neonatal lethality, mainly due to the cannibalistic nature of rodents in response to pups that are not thriving or are ill. Cannibalism can be so extensive that carcass remnants are not evident in the cage making it difficult to determine if a litter was actually born. Frequent observation of pups (taking care not to overly disturb the dam) for transgenic lines that are exhibiting breeding or production problems is imperative to determine if pups are not born at all due to embryo death, retention, and resorption, if pups are born dead, or if pups are born live and then die. With embryonic or perinatal lethal recessives, litter sizes will be reduced and an absence or dramatic reduction of homozygote genotypes will be discernible across multiple litters from different breeders. Definitive determination of embryonic lethality may result in the need to sacrifice pregnant females to determine embryonic viability and to harvest embryos for genotyping. In the case of a proven embryonic lethal phenotype, re-creation of the model with an inducible expression system, such as a tet-on system, will need to be considered for further study and production of viable homozygote offspring. True perinatal or neonatal lethality can be difficult to prove. Deficiencies in maternal rearing capabilities and any adverse impact of the micro- or macro-environment must first be eliminated. Once these potential causes of neonatal mortality have been eliminated, pups should be closely monitored daily for clinical abnormalities and viability to determine the age range where mortality occurs. Pathologic assessment of pups before the determined age interval is then advised.

Some genetic alterations have direct effects on fertility. Such effects can impact copulatory behavior,[206] sperm motility and morphology,[207] fertilization, uterine implantation,[208] or ability to care for pups.[209] These phenotypic effects may be predicted, based on the function of the gene in humans, or unexpected, leading to a thorough investigation. Documentation of reproductive performance allows for the rapid identification of these phenotypes when they occur.

Other factors

Other factors that may affect colony performance:

Stress	Illness
Phenotype	Strain background
Diet	Light
Noise and vibration	Staff
Housing	Temperature
Seasonal changes	Odors (human or rodent)

In addition to the factors discussed previously, a host of other things may also influence breeding performance. An important fact to remember is that reproduction is a luxury function. When an animal is stressed, it is more concerned with remaining alive. Ensuring its future through genetic contribution to the next generation is hardly possible if the animal is dead. If animals are stressed, for whatever reason, reproduction will suffer.

The nutritional plane of the breeders can influence productivity of the animals. An obese male is not able to execute normal mounting behavior and intromissions. He may have a low libido or preputial infections because he cannot groom effectively. His sperm may also be of poor quality, resulting in subfertility.[210,211] An obese female may not have the hormonal milieu to support ovulation, let alone pregnancy, depending on the level of obesity.[212] She may also have difficulties during parturition. Obese and underweight females may not be able to nutritionally support lactating pups. Rodents are some of the few animals produced in large numbers that are not given nutritional support for breeding and lactation. When considering diet, it is good to consider that animals may benefit from either an increase or decrease in fat. The literature supports that different strains have different macronutrient preferences, and milk quality has been shown to be affected by the type of fat in the diet.[213-215]

The health of an individual animal can have an influence on breeding performance. Overall colony health can also affect breeding performance, but agents currently prevalent in rodent facilities tend to have effects on research rather than animal health. If animals cannot breathe due to overwhelming *Pneumocystis* infection, they certainly will not breed. If a clinically ill rodent does breed, it is at risk for fetal death and resorption.

Changes in the environment often result in production problems that extend across multiple transgenic lines and sometimes, across multiple rodent rooms within a facility. Disruption of the light cycle can have a profound impact on successful breeding and normal animal behavior.[216] The usual problem is persistent light, as a failure of the lights to turn on is detected during normal working hours. Most breeding behavior takes place during the first few hours of the dark cycle. This is unfortunately, the most frequent time that dark cycles are interrupted by manual light override. Regular manual light override will result in a decrease in the number of pregnancies detected and litters produced. Failure of automated light systems to switch off, resulting in 24 hours of light, will cause persistent estrus in rats.[217] Although wild mice are usually described as "long-day breeders," in many cases this is more truly related to resource availability rather than day length.[218] Many facilities use a 14:10 light:dark cycle to attempt to positively influence mouse reproduction.

In the authors' experience, this light cycle neither helps nor harms rat reproduction. In addition, rodents see into the UV spectrum, but full-spectrum lights are rarely, if ever, used in rodent facilities.

Noise and vibration can also have adverse effects on pup viability with a noted increase in stillborn pups cannibalism in cages with pups less than one week of age.[219] Noise may also result in seizures in some strains, such as DBA/2 and FVB/N. Again, in the authors' experience, repeated, constant, or regular noises allow the animals to adapt somewhat (rats and mice breed well in subways, for example), but intermittent or unpredictable noises can have negative effects. Noise and vibration as a result of nearby blasting for construction or use of an air drill or jackhammer during renovations can devastate litters in close proximity. Vibrations within the animal facility as a result of normal, regular activities also need to be considered as having potential to adversely affect production and pup viability. Vibrations within walls from HVAC units, in individually ventilated racks, and as a result of rack and cage movement for cleaning may need to be addressed when attempting to minimize vibrations that might affect fragile colonies. Consideration should be given, however, to the hearing range of mice and rats and if nearby activities are audible to them.[220] For ultrasonic emissions, commercially available bat detectors can shift the sound to human hearing range.

Environmental temperature within the room, as well as within the cage, must be considered in establishing optimal breeding conditions. Temperature is influenced by air movement, and many animals are housed in individually ventilated cages. Some types of individually ventilated cages have air moving through them at rapid speed from small apertures, a situation that many mice may find both aversive and chilly.
The thermoneutral zone of mice is from 29.6°C to 30.5°C, however, 21°C - 25°C is comfortable for people working in protective equipment. Provision of nesting material allows the animals control over their environment as well as expression of normal behavior. Nesting material or shelters also allow

dams to keep neonates warm, thus improving pup survival. The use and function of pheromones for mice and rats has been discussed and it is not surprising that odors can affect rodent breeding behavior and maternal care. Scent is one of the mouse's primary sensory modalities, but often little thought is given to managing pheromones in colonies. For most mice, this will not matter, but for sensitive strains, it can be very important. Never house breeding cages adjacent to each other, or adjacent to any cage containing a sexually mature male. Avoid all possible contact with odor from bedding from breeding or male cages. This requires careful monitoring of, and a little extra effort with, cage changing by making sure the dirty litter bin is on the opposite side of the room from breeders, for example. Harem systems can be challenging for some breeding mice, since dominant females can suppress the reproductive maturity of subordinates in the cage. Try to keep the male in the cage because male pheromones accelerate reproductive maturity in female offspring. Odors from disinfectants or chemicals associated with animal facility renovation may have an impact on fertility, pheromonal recognition leading to copulatory behavior, and pheromonal recognition of pups by the dam.[221] When investigating production-related problems across multiple rooms within a facility, consider any recent changes to animal care and husbandry practices. Additionally, the introduction of new animal care or research staff into a breeding rodent room may have a negative effect on production parameters if they are new to rodent handling or wear strong scents. Training for these individuals and explanation about why these scents are a problem is important to avoid these influences.

As tightly as the animals' environment is controlled today, changes in production parameters are still evident as a result of seasonal changes.[222,223] A seasonal breeding depression is often noted during late fall and winter months. There is no remedy for this phenomenon but advance planning to increase the number of breeder pairs set up will permit breeding goals to be met.

Solutions

Once the problem has been identified and investigated, change can be initiated. Solutions should be thoroughly discussed and documented to enable recognition of which remedy was pivotal in solving a production-related problem. If animals have breeding problems, documentation of their special needs should accompany them if they are shared with other investigators. Some solutions for rescue of a transgenic line can be costly and can also require euthanasia of valuable breeders in the case of assisted reproductive technologies. Simple, easy to implement changes should be attempted initially before moving on to more complicated fixes.
If animals are not breeding successfully, establishing new breeder pairs will often "kick-start" breeding. Additionally, introduction of new breeding stock, such as young, breeding age, inbred females, to established breeder males may also help. If these two approaches fail, use of a superovulation hormone protocol (HCG and PMSG) to synchronize the estrus cycle of the females is suggested. In older females, this protocol will not result in overproduction of oocytes but will serve to artificially initiate the gonadal-hypothalamic-pituitary axis, bringing females into sexual receptivity.
During all of these efforts, it is advised to time the mating of breeders and check the females daily, first thing in the morning, for plugs that would indicate successful breeding overnight. If necessary, a timed breeding paradigm can be used but recognize that effort, time, and skill are required to successfully accomplish this. If a limited number of females are available, daily vaginal swabs must be collected and read to determine if females are cycling regularly and when to place a female into a male's cage for breeding. Checking for copulatory plugs is also required to confirm that breeding has been successful. This approach is also sometimes used to track the specific day of gestation for embryonic development studies.

If the problem is related to maternal care, pup fostering should be considered. Again, this paradigm is labor intensive and requires additional animals to serve as foster dams. Fostering is used for lactational problems and for pups that are not thriving due to deficient maternal rearing capabilities. Use of outbred or hybrid rodents as foster dams is ideal due to their excellent mothering abilities. These foster dams must have a coat color different than the fosterlings. Knowledge of gestation lengths for the foster dam as well as the donor dam is critical because the gestation length of the foster female may be different than that of the transgenic female and the births of both litters must be timed to have pups as close in age as possible. Pups from each litter should ideally be within one to two days of age of each other. Pups can be fostered to dams with older litters (up to 5 days older) if that is all that is available. The number of fosterlings should approximate that of the original foster litter, with foster pups euthanized to adjust the total litter size. There are many specific techniques to actually perform the fostering and is it often the preference of the individual managing the fostering program in choosing that technique.

The authors prefer the following technique:
Gently remove the foster mom from the home cage. take the pups that will be euthanized and set them to the side. Place the new pups into the foster nest and gently mix the new pups with any remaining pups used to equilibrate the litter size. Rub the nesting material and dirty bedding for the foster cage over the new pups. Pick up the foster mom by the tail or by scruffing and hold her over the combined litter, stimulating the foster mom to urinate on the litter. Then, place the foster mom into the cage and replace the cage on the rack. When monitoring the foster mom for acceptance of the new pups, do not disturb the cage and its inhabitants.

Other assisted reproductive techniques (ART), such as in vitro fertilization[224,225] and intracytoplasmic sperm injection,[226,227] are well-known techniques in mice and rats, but will not be described further. ARTs are generally a last resort for a strain, or used to obtain a large number of animals very rapidly. They also have their place in cryopreservation[228] and embryo transfer rederivation of colonies.

Conclusion

Breeding rodents for research use can be a challenging process with many pitfalls. It is a multi-step process with challenges possible at many different points in the cycle. Environment, genetics, husbandry can all affect breeding and colony performance.

If you breed rodents, share information with colleagues and facility veterinarians. Stay organized so that problems will be noticed quickly. If a problem is suspected, be proactive. It is unlikely problems will get better on their own, so help is always advisable, even if it is just brainstorming with someone about what might be wrong. The professionals at Charles River are ready to help you with your colony management challenges at any time.

References

1 Home Office. Statistics of Scientific Procedures on Living Animals: Great Britain 2010. (The Stationery Office, 2011).

2 Waterston, R. H. *et al*. Initial sequencing and comparative analysis of the mouse genome. *Nature* **420**, 520-562 (2002).

3 Buehr, M. *et al*. Capture of authentic embryonic stem cells from rat blastocysts. *Cell* **135**, 1287-1298 (2008).

4 Geurts, A. M. et al. Knockout rats via embryo microinjection of zinc-finger nucleases. *Science* **325**, 433 (2009).

5 Voigt, B. & Serikawa, T. Pluripotent stem cells and other technologies will eventually open the door for straightforward gene targeting in the rat. *Dis Model Mech* **2**, 341-343 (2009).

6 Silver, L. M. *Mouse Genetics: Concepts and Applications*. 1st edn, 3-31 (Oxford University Press, 1995).

7 Hedrich, H. J. in *The Laboratory Rat* (ed Georg J. Krinke) 3-16 (Academic Press, 2000).

8 Keeler, C. E. *The Laboratory Mouse: Its Origin, Heredity, and Culture*. 7-18 (Harvard University Press, 1931).

9 Lang, A. *Custom and Myth*. (Longmans, Green, and Co., 1893).

10 Keeler, C. E. & Fuji, S. The antiquity of mouse variations in the Orient. *Journal of Heredity* **28**, 93-96 (1937).

11 Festing, M. F. W. & Lovell, D. P. in *Symposium of the Zoological Society of London, No. 47: Biology of the house mouse* (ed R. J. Berry) 43-62 (Academic Press, 1981).

12 Lathrop, A. E. & Loeb, L. Further investigations on the origin of tumors in mice : I. Tumor incidence and tumor age in various strains of mice. *J Exp Med* **22**, 646-673 (1915).

13 Richter, C. P. The effects of domestication and selection on the behavior of the Norway rat. *Journal of the National Cancer Institute* **15**, 727-738 (1954).

14 Lindsey, J. R. & Baker, H. J. in *The Laboratory Rat* (eds Mark A. Suckow, Steven H. Weisbroth, & C. L. Franklin) 1-52 (Academic Press, 2006).

15 Calhoun, J. B. *The Ecology and Sociology of the Norway Rat*. (US Public Health Service, 1963).

16 Crowcroft, P. *Mice All Over*. (Chicago Zoological Society, 1966).

17 Gray, S. J., Jensen, S. P. & Hurst, J. L. Structural complexity of territories: preference, use of space and defence in commensal house mice, *Mus domesticus*. *Anim Behav* **60**, 765-772 (2000).

18 Arakawa, H., Blanchard, D. C. & Blanchard, R. J. Colony formation of C57BL/6J mice in visible burrow system: identification of eusocial behaviors in a background strain for genetic animal models of autism. *Behav Brain Res* **176**, 27-39 (2007).

19 Blanchard, R. J. & Blanchard, D. C. Antipredator defensive behaviors in a visible burrow system. *J Comp Psychol* **103**, 70-82 (1989).

20 Baker, A. E. M. Gene flow in house mice: Behavior in a population cage. *Behavioral Ecology and Sociobiology* **8**, 83-90 (1981).

21 Drickamer, L. C., Gowaty, P. A. & Wagner, D. M. Free mutual mate preferences in house mice affect reproductive success and offspring performance. *Animal Behaviour* **65**, 105-114 (2003).

22 Yamazaki, K. & Beauchamp, G. K. Genetic basis for MHC-dependent mate choice. *Adv Genet* **59**, 129-145 (2007).

23 Haemisch, A. & Gartner, K. The cage design affects intermale aggression in small groups of male laboratory mice: strain specific consequences on social organization, and endocrine activations in two inbred strains (DBA/2J and CBA/J). *Journal of Experimental Animal Science* **36**, 101-116 (1994).

24 Nevison, C. M., Hurst, J. L. & Barnard, C. J. Strain-specific effects of cage enrichment in male laboratory mice (*Mus musculus*). *Animal Welfare* **8**, 361-379 (1999).

25 Van Loo, P. L. et al. Influence of cage enrichment on aggressive behaviour and physiological parameters in male mice. *Applied Animal Behaviour Science* **76**, 65-81 (2002).

26 Van Loo, P. L., Van Zutphen, L. F. & Baumans, V. Male management: Coping with aggression problems in male laboratory mice. *Lab Anim* **37**, 300-313 (2003).

27 Howerton, C. L., Garner, J. P. & Mench, J. A. Effects of a running wheel-igloo enrichment on aggression, hierarchy linearity, and stereotypy in group-housed male CD-1 (ICR) mice. *Applied Animal Behaviour Science* **115**, 90-103 (2008).

28 Kaliste, E. K., Mering, S. M. & Huuskonen, H. K. Environmental modification and agonistic behavior in NIH/S male mice: nesting material enhances fighting but shelters prevent it. *Comp Med* **56**, 202-208 (2006).

29 Berridge, K. C., Fentress, J. C. & Parr, H. Natural syntax rules control action sequence of rats. *Behav Brain Res* **23**, 59-68 (1987).

30 Fentress, J. C. Expressive contexts, fine structure, and central mediation of rodent grooming. *Ann N Y Acad Sci* **525**, 18-26 (1988).

31 Garner, J. P., Weisker, S. M., Dufour, B. & Mench, J. A. Barbering (fur and whisker trimming) by laboratory mice as a model of human trichotillomania and obsessive-compulsive spectrum disorders. *Comp Med* **54**, 216-224 (2004).

32 Garner, J. P., Dufour, B., Gregg, L. E., Weisker, S. M. & Mench, J. A. Social and husbandry factors affecting the prevalence and severity of barbering ('whisker trimming') by laboratory mice. *Applied Animal Behaviour Science* **89**, 263-282 (2004).

33 Nicholson, A. et al. The response of C57BL/6J and BALB/cJ mice to increased housing density. *J Am Assoc Lab Anim Sci* **48**, 740-753 (2009).

34 Kalueff, A. V., Minasyan, A., Keisala, T., Shah, Z. H. & Tuohimaa, P. Hair barbering in mice: implications for neurobehavioural research. *Behav Processes* **71**, 8-15 (2006).

35 Durham, D. & Woolsey, T. A. Acute whisker removal reduces neuronal activity in barrels of mouse SmL cortex. *J Comp Neurol* **178**, 629-644 (1978).

36 Bechard, A., Meagher, R. & Mason, G. Environmental enrichment reduces the likelihood of alopecia in adult C57BL/6J mice. *J Am Assoc Lab Anim Sci* **50**, 171-174 (2011).

37 Dufour, B. D. *et al*. Nutritional up-regulation of serotonin paradoxically induces compulsive behavior. *Nutr Neurosci* **13**, 256-264 (2010).

38 Van Loo, P. L. & Baumans, V. The importance of learning young: the use of nesting material in laboratory rats. *Lab Anim* **38**, 17-24 (2004).

39 Gaskill, B. N., Rohr, S. A., Pajor, E. A., Lucas, J. R. & Garner, J. P. Some like it hot: Mouse temperature preferences in laboratory housing *Applied Animal Behaviour Science* **116**, 279-285 (2009).

40 Hess, S. E. *et al*. Home improvement: C57BL/6J mice given more naturalistic nesting materials build better nests. *J Am Assoc Lab Anim Sci* **47**, 25-31 (2008).

41 Vitalo, A. *et al.* Nest making and oxytocin comparably promote wound healing in isolation reared rats. *PLoS One* **4**, e5523 DOI 5510.1371/journal.pone.0005523 (2009).

42 Tilly, S. C., Dallaire, J. & Mason, G. J. Middle-aged mice with enrichment-resistant stereotypic behaviour show reduced motivation for enrichment. *Animal Behaviour* **80**, 363-373 (2010).

43 Lathe, R. The individuality of mice. *Genes, brain, and behavior* **3**, 317-327 (2004).

44 Oliva, A. M. *et al*. Toward a mouse neuroethology in the laboratory environment. *PLoS One* **5**, e11359 DOI 11310.11371/journal.pone.0011359 (2010).

45 Hutchinson, E., Avery, A. & Vandewoude, S. Environmental enrichment for laboratory rodents. *ILAR J* **46**, 148-161 (2005).

46 Simpson, J. & Kelly, J. P. The impact of environmental enrichment in laboratory rats--behavioural and neurochemical aspects. *Behav Brain Res* **222**, 246-264 (2011).

47 Gonder, J. C. & Laber, K. A renewed look at laboratory rodent housing and management. *ILAR J* **48**, 29-36 (2007).

48 Latham, N. & Mason, G. From house mouse to mouse house: the behavioural biology of free-living *Mus musculus* and its implications in the laboratory. *Applied Animal Behaviour Science* **86**, 261-289 (2004).

49 Panksepp, J. Neuroevolutionary sources of laughter and social joy: modeling primal human laughter in laboratory rats. *Behav Brain Res* **182**, 231-244 (2007).

50 Hurst, J. L. & West, R. S. Taming anxiety in laboratory mice. *Nat Methods* **7**, 825-826 (2010).

51 Gordon, C. J. *Temperature Regulation in Laboratory Rodents.* (Cambridge University Press, 1993).

52 Gordon, C. J. Effect of cage bedding on temperature regulation and metabolism of group-housed female mice. *Comp Med* **54**, 63-68 (2004).

53 Le, N.-M. & Brown, J. W. Characterization of the thermoneutral zone of the laboratory rat. *FASEB Journal* **22**, 956.919 (2008).

54 Flower, D. R. The lipocalin protein family: structure and function. *Biochem J* **318** (Pt 1), 1-14 (1996).

55 Zegarelli, E. V. Adamantoblastomas in the Slye stock of mice. *American Journal of Pathology* **20**, 23-87 (1944).

56 Weinreb, M. M., Assif, D. & Michaeli, Y. Role of attrition in the physiology of the rat incisor. I. the relative value of different components of attrition and their effect on eruption. *Journal of Dental Research* **46**, 527-531 (1967).

57 Ebino, K. Y. Studies on coprophagy in experimental animals. *Jikken Dobutsu* **42**, 1-9 (1993).

58 Brown, R. E. & Wong, A. A. The influence of visual ability on learning and memory performance in 13 strains of mice. *Learn Mem* **14**, 134-144 (2007).

59 Wong, A. A. & Brown, R. E. Visual detection, pattern discrimination and visual acuity in 14 strains of mice. *Genes, Brain, and Behavior* **5**, 389-403 (2006).

60 Prusky, G. T., Harker, K. T., Douglas, R. M. & Whishaw, I. Q. Variation in visual acuity within pigmented, and between pigmented and albino rat strains. *Behav Brain Res* **136**, 339-348 (2002).

61 Amleh, A. & Dean, J. Mouse genetics provides insight into folliculogenesis, fertilization and early embryonic development. Hum. *Reprod Update.* **8**, 395-403 (2002).

62 Serfilippi, L. M., Pallman, D. R., Gruebbel, M. M., Kern, T. J. & Spainhour, C. B. Assessment of retinal degeneration in outbred albino mice. *Comp Med* **54**, 69-76 (2004).

63 Clapcote, S. J., Lazar, N. L., Bechard, A. R., Wood, G. A. & Roder, J. C. NIH Swiss and Black Swiss mice have retinal degeneration and performance deficits in cognitive tests. *Comp Med* **55**, 310-316 (2005).

64 LaVail, M. M., Sidman, R. L. & Gerhardt, C. O. Congenic strains of RCS rats with inherited retinal dystrophy. *J Hered* **66**, 242-244 (1975).

65 Balkema, G. W. & Drager, U. C. Impaired visual thresholds in hypopigmented animals. *Vis Neurosci* **6**, 577-585 (1991).

66 Pecci Saavedra, J. & Pellegrino de Iraldi, A. Retinal alterations induced by continuous light in immature rats. I. Fine structure and electroretinography. *Cell Tissue Res* **166**, 201-211 (1976).

67 Heffner, H. E. & Heffner, R. S. Hearing ranges of laboratory animals. *J Am Assoc Lab Anim Sci* **46**, 20-22 (2007).

68 Hahn, M. E. & Lavooy, M. J. A review of the methods of studies on infant ultrasound production and maternal retrieval in small rodents. *Behav Genet* **35**, 31-52 (2005).

69 Panksepp, J. & Burgdorf, J. 50-kHz chirping (laughter?) in response to conditioned and unconditioned tickle-induced reward in rats: effects of social housing and genetic variables. *Behav Brain Res* **115**, 25-38 (2000).

70 Holy, T. E. & Guo, Z. Ultrasonic songs of male mice. *PLoS Biol* **3**, e386 DOI 310.1371/journal.pbio.0030386 (2005).

71 Musolf, K., Hoffman, F. & Penn, D. J. Ultrasonic courtship vocalizations in wild house mice, *Mus musculus musculus. Animal Behavior* **79**, 757-764 (2010).

72 Johnson, K. R., Zheng, Q. Y. & Erway, L. C. A major gene affecting age-related hearing loss is common to at least ten inbred strains of mice. *Genomics* **70**, 171-180 (2000).

73 Burn, C. C., Peters, A., Day, M. J. & Mason, G. J. Long-term effects of cage-cleaning frequency and bedding type on laboratory rat health, welfare, and handleability: a cross-laboratory study. *Laboratory Animals* **40**, 353-370 (2006).

74 Van Loo, P. L., Kruitwagen, C. L. J. J., Van Zutphen, B. F., Koolhaas, J. M. & Baumans, V. Modulation of aggression in male mice: influence of cage cleaning regime and scent marks. *Animal Welfare* **9**, 281-295 (2000).

75 Koltzenburg, M., Stucky, C. L. & Lewin, G. R. Receptive properties of mouse sensory neurons innervating hairy skin. *J Neurophysiol* **78**, 1841-1850 (1997).

76 Berg, R. W. & Kleinfeld, D. Rhythmic whisking by rat: retraction as well as protraction of the vibrissae is under active muscular control. *J Neurophysiol* **89**, 104-117 (2003).

77 Carvell, G. E. & Simons, D. J. Biometric analyses of vibrissal tactile discrimination in the rat. *J Neurosci* **10**, 2638-2648 (1990).

78 Cybulska-Klosowicz, A. & Kossut, M. Mice can learn roughness discrimination with vibrissae in a jump stand apparatus. *Acta Neurobiol Exp (Wars)* **61**, 73-76 (2001).

79 Pritchett, K. R. & Taft, R. A. in *The Mouse in Biomedical Research: Normative Biology, Husbandry, and Models* Vol. 3 *The Mouse in Biomedical Research* (eds *J.* Fox *et al.*) Ch. 3, 91-122 (Academic Press, 2007).

80 Lohmiller, J. J. & Swing, S. P. in *The Laboratory Rat* (eds Mark A. Suckow, Steven H. Weisbroth, & C. L. Franklin) 147-164 (Academic Press, 2006).

81 Hardy, P. in *The Handbook of Experimental Animals: The Laboratory Mouse* (eds Hans J. Hedrich & Gillian Bullock) 409-434 (Elsevier, 2004).

82 Maeda, K., Ohkura, S. & Tsukamura, H. in *The Handbook of Experimental Animals: The Laboratory Rat* (ed Georg G. Krinke) 145-176 (Academic Press, 2000).

83 Zimmerman, F., Weiss, J. & Reifenberg, K. in *The Handbook of Experimental Animals: The Laboratory Rat* (ed Georg G. Krinke) 177-198 (Academic Press, 2000).

84 Ruwanpura, S. M., McLachlan, R. I. & Meachem, S. J. Hormonal regulation of male germ cell development. *J Endocrinol* **205**, 117-131 (2010).

85 Honma, S. et al. Low dose effect of in utero exposure to bisphenol A and diethylstilbestrol on female mouse reproduction. *Reprod Toxicol* **16**, 117-122 (2002).

86 Hotchkiss, A. K. & Vandenbergh, J. G. The anogenital distance index of mice (*Mus musculus domesticus*): an analysis. *Contemp Top Lab Anim Sci* **44**, 46-48 (2005).

87 Ryan, B. C. & Vandenbergh, J. G. Intrauterine position effects. *Neurosci Biobehav Rev* **26**, 665-678 (2002).

88 Dunbar, M. E. *et al*. Parathyroid hormone-related protein signaling is necessary for sexual dimorphism during embryonic mammary development. *Development* **126**, 3485-3493 (1999).

89 Roosen-Runge, E. C. Process of spermatogenesis in mammals. *Biol Rev* **37**, 343-377 (1962).

90 Sztein, J. M., Farley, J. S. & Mobraaten, L. E. *In vitro* fertilization with cryopreserved inbred mouse sperm. *Biol Reprod* **63**, 1774-1780 (2000).

91 McGill, T. E. & Blight, W. C. Effects of genotype on the recovery of the sex drive in the male mouse. *J Comp Physiol Psychol* **56**, 887-888 (1963).

92 Karen, L. M. & Barfield, R. J. Differential rates of exhaustion and recovery of several parameters of male rat sexual behavior. *J Comp Physiol Psychol* **88**, 693-703 (1975).

93 Dean, M. D., Ardlie, K. G. & Nachman, M. W. The frequency of multiple paternity suggests that sperm competition is common in house mice (*Mus domesticus*). *Mol Ecol* **15**, 4141-4151 (2006).

94 Szabo, K. T., Free, S. M., Birkhead, H. A. & Gay, P. E. Predictability of pregnancy from various signs of mating in mice and rats. *Lab Anim Care* **19**, 822-825 (1969).

95 Parkes, A. S. The length of the oestrous cycle in the unmated normal mouse: records of one thousand cycles. *Br J Exp Biol* **5**, 371-377 (1928).

96 Champlin, A. K., Dorr, D. L. & Gates, A. H. Determining the stage of the estrous cycle in the mouse by the appearance of the vagina. *Biology of Reproduction* **8**, 491-494 (1973).

97 Nelson, J. F., Felicio, L. S., Randall, P. K., Sims, C. & Finch, C. E. A longitudinal study of estrous cyclicity in aging C57BL/6J mice: I. Cycle frequency, length and vaginal cytology. *Biol Reprod* **27**, 327-339 (1982).

98 Marcondes, F. K., Bianchi, F. J. & Tanno, A. P. Determination of the estrous cycle phases of rats: some helpful considerations. *Braz J Biol* **62**, 609-614 (2002).

99 Runner, M. N. & Ladman, A. J. The time of ovulation and its diurnal regulation in the post-parturitional mouse. *Anat Rec* **108**, 343-361 (1950).

100 Bingel, A. S. Further studies of post-partum ovulation timing in mice. *J Reprod Fertil* **65**, 313-318 (1982).

101 Gilbert, A. N., Rosenwasser, A. M. & Adler, N. T. Timing of parturition and postpartum mating in Norway rats: interaction of an interval timer and a circadian gate. *Physiol Behav* **34**, 61-63 (1985).

102 Lopes, F. L., Desmarais, J. A. & Murphy, B. D. Embryonic diapause and its regulation. *Reproduction* **128**, 669-678 (2004).

103 Renfree, M. B. & Shaw, G. Diapause. *Annu Rev Physiol* **62**, 353-375 (2000).

104 Byers, S. L., Payson, S. J. & Taft, R. A. Performance of ten inbred mouse strains following assisted reproductive technologies (ARTs). *Theriogenology* **65**, 1716-1726 (2006).

105 Popova, E., Bader, M. & Krivokharchenko, A. Strain differences in superovulatory response, embryo development and efficiency of transgenic rat production. *Transgenic Res* **14**, 729-738 (2005).

106 Marsden, H. M. & Bronson, F. H. The synchrony of oestrus in mice: relative roles of the male and female environments. *J Endocrin* **32**, 313-319 (1965).
107 McClintock, M. K. Estrous synchrony: modulation of ovarian cycle length by female pheromones. *Physiol Behav* **32**, 701-705 (1984).
108 van der Lee, S. & Boot, L. M. Spontaneous pseudopregnancy in mice. *Acta Physiol Pharmacol* **4**, 442-444 (1955).
109 Whitten, W. K. Modification of the oestrous cycle of the mouse by external stimuli associated with the male. *J Endocrinol* **13**, 399-404 (1956).
110 Bruce, H. M. A block to pregnancy in the mouse caused by proximity of strange males. *J Reprod Fert* **1**, 96-103 (1960).
111 Parkes, A. S. & Bruce, H. M. Olfactory stimuli in mammalian reproduction. *Science* **134**, 1049-1054 (1961).
112 Peele, P., Salazar, I., Mimmack, M., Keverne, E. B. & Brennan, P. A. Low molecular weight constituents of male mouse urine mediate the pregnancy block effect and convey information about the identity of the mating male. *Eur.J Neurosci.* **18**, 622-628 (2003).
113 Hurst, J. L. Female recognition and assessment of males through scent. *Behav Brain Res* **200**, 295-303 (2009).
114 McGill, T. E. Sexual behavior in three inbred strains of mice. *Behaviour* **19**, 341-350 (1962).
115 Bennett, J. P. & Vickery, B. H. in *Reproduction and Breeding Techniques for Laboratory Animals* (ed E. S. E. Hafez) 299-315 (Lea and Febiger, 1970).
116 Diakow, C. A. Motion picture analysis of rat mating behavior. *Journal of Comparative and Physiological Psychology* **88**, 704-712 (1975).
117 Murray, S. A. *et al.* Mouse gestation length is genetically determined. *PLoS One* **5**, e12418 DOI 12410.11371/journal.pone.0012418 (2010).
118 McLaren, A. & Michie, D. Nature of the systemic effect of litter size on gestation period in mice. *J Reprod Fertil* **6**, 139-141 (1963).
119 Weber, E. M. & Olsson, I. A. S. Maternal behaviour in *Mus musculus* sp.: An ethological review. *Applied Animal Behaviour Science* **114**, 1-22 (2008).
120 Flint, A. P., Heap, R. B., Ingram, D. L. & Walters, D. E. The effect of day length on the duration of pregnancy and the onset of parturition in the rat. *Q J Exp Physiol* **71**, 285-293 (1986).
121 Berry, R. J. The natural history of the house mouse. *Field Studies* **3**, 219-262 (1970).
122 Fuller, G. B., McGee, G. E., Nelson, J. C., Willis, D. C. & Culpepper, R. D. Birth sequence in mice. *Lab Anim Sci* **26**, 198-200 (1976).
123 Labov, J. B. Factors influencing infanticidal behavior in wild male house mice (*Mus musculus*). *Behavioral Ecology and Sociobiology* **6**, 297-303 (1980).
124 Perrigo, G., Belvin, L. & Vom Saal, F. S. Time and sex in the male mouse: temporal regulation of infanticide and parental behavior. *Chronobiol Int* **9**, 421-433 (1992).
125 Mennella, J. A. & Moltz, H. Pheromonal emission by pregnant rats protects against infanticide by nulliparous conspecifics. *Physiol Behav* **46**, 591-595 (1989).

126 Mennella, J. A. & Moltz, H. Infanticide in rats: male strategy and female counter-strategy. *Physiol Behav* **42**, 19-28 (1988).
127 Jakubowski, M. & Terkel, J. Transition from pup killing to parental behavior in male and virgin female albino rats. *Physiol Behav* **34**, 683-686 (1985).
128 Mann, M. A., Kinsley, C., Broida, J. & Svare, B. Infanticide exhibited by female mice: genetic, developmental and hormonal influences. *Physiol Behav* **30**, 697-702 (1983).
129 Shoji, H. & Kato, K. Maternal behavior of primiparous females in inbred strains of mice: a detailed descriptive analysis. *Physiol Behav* **89**, 320-328 (2006).
130 Branchi, I., Santucci, D. & Alleva, E. Ultrasonic vocalisation emitted by infant rodents: a tool for assessment of neurobehavioural development. *Behavioural Brain Research* **125**, 49-56 (2001).
131 Russell, J. A. Milk yield, suckling behaviour and milk ejection in the lactating rat nursing litters of different sizes. *J Physiol* **303**, 403-415 (1980).
132 Knight, C. H., Maltz, E. & Docherty, A. H. Milk yield and composition in mice: effects of litter size and lactation number. *Comp Biochem Physiol A Comp Physiol* **84**, 127-133 (1986).
133 Hayes, L. D. To nest communally or not to nest communally: a review of rodent communal nesting and nursing. *Anim Behav* **59**, 677-688 (2000).
134 Manning, C. J., Dewsbury, D. A., Wakeland, E. K. & Potts, W. K. Communal nesting and communal nursing in house mice, *Mus musculus domesticus. Animal Behaviour* **50**, 741-751 (1995).
135 Albert, M. & Roussel, C. Changes from puberty to adulthood in the concentration, motility and morphology of mouse epididymal spermatozoa. *Int J Androl* **6**, 446-460 (1983).
136 Robb, G. W., Amann, R. P. & Killian, G. J. Daily sperm production and epididymal sperm reserves of pubertal and adult rats. *J Reprod Fertil* **54**, 103-107 (1978).
137 Schmidt, J. A., Oatley, J. M. & Brinster, R. L. Female mice delay reproductive aging in males. *Biol Reprod* **80**, 1009-1014 (2009).
138 Wang, C., Leung, A. & Sinha-Hikim, A. P. Reproductive aging in the male brown-Norway rat: a model for the human. *Endocrinology* **133**, 2773-2781 (1993).
139 Matt, D. W., Sarver, P. L. & Lu, J. K. Relation of parity and estrous cyclicity to the biology of pregnancy in aging female rats. *Biol Reprod* **37**, 421-430 (1987).
140 Biggers, J. D., Finn, C. A. & McLaren, A. Long-term reproduction performance of female mice II. variation of litter size with parity. *J Reprod Fert* **3**, 313-330 (1962).
141 Weiling, F. Historical study: Johann Gregor Mendel 1822-1884. *Am J Med Genet* **40**, 1-25; discussion 26 (1991).
142 Silvers, W. K. *The Coat Colors of Mice: A model for mammalian gene action and interaction*. (Springer-Verlag, 1979).
143 Nagy, A., Gertsenstein, M., Vintersten, K. & Behringer, R. *Manipulating the Mouse Embryo: A Laboratory Manual*. 3rd edn, (Cold Spring Harbor Laboratory Press, 2002).

144 Pinkert, C. A. Transgenic Animal Technology: A Laboratory Handbook. 2nd edn, (Academic Press, 2002).

145 Cartwright, E. J. Transgenesis Techniques: Principles and Protocols. 3rd edn, *Methods in Molecular Biology* **348** p. (Humana Press, 2009).

146 Kühn, R. & Wurst, W. Gene Knockout Protocols. 2nd edn, *Methods in Molecular Biology* **532** p. (Humana Press, 2009).

147 Joyner, A. L. Gene Targeting: A Practical Approach. *The Practical Approach Series* **293** p. (Oxford University Press, 2000).

148 Gordon, J. W., Scangos, G. A., Plotkin, D. J., Barbosa, J. A. & Ruddle, F. H. Genetic transformation of mouse embryos by microinjection of purified DNA. *Proc Natl Acad Sci U S A* **77**, 7380-7384 (1980).

149 Chung, S. *et al*. Analysis of different promoter systems for efficient transgene expression in mouse embryonic stem cell lines. Stem Cells 20, 139-145 (2002).

150 Probst, F. J. *et al*. Correction of deafness in shaker-2 mice by an unconventional myosin in a BAC transgene. *Science* **280**, 1444-1447 (1998).

151 Gossen, M. & Bujard, H. Tight control of gene expression in mammalian cells by tetracycline-responsive promoters. *Proc Natl Acad Sci U S A* **89**, 5547-5551 (1992).

152 Macarthur, C. *et al*. Chromatin insulator elements block transgene silencing in engineered hESC lines at a defined chromosome 13 locus. *Stem Cells Dev* DOI 10.1089/scd.2011.0163 (2011).

153 Doyle, A., McGarry, M. P., Lee, N. A. & Lee, J. J. The construction of transgenic and gene knockout/knockin mouse models of human disease. *Transgenic Res* DOI 10.1007/s11248-11011-19537- 11243 (2011).

154 Thomas, K. R. & Capecchi, M. R. Site-directed mutagenesis by gene targeting in mouse embryo-derived stem cells. *Cell* **51**, 503-512 (1987).

155 Doetschman, T. C. Gene targeting in embryonic stem cells. *Biotechnology* **16**, 89-101 (1991).

156 Martin, G. R. Isolation of a pluripotent cell line from early mouse embryos cultured in medium conditioned by teratocarcinoma stem cells. *Proc Natl Acad Sci U S A* **78**, 7634-7638 (1981).

157 Evans, M. J. & Kaufman, M. H. Establishment in culture of pluripotential cells from mouse embryos. *Nature* **292**, 154-156 (1981).

158 Martin, G. R., Silver, L. M., Fox, H. S. & Joyner, A. L. Establishment of embryonic stem cell lines from preimplantation mouse embryos homozygous for lethal mutations in the t-complex. *Dev Biol* **121**, 20-28 (1987).

159 Kawase, E. *et al*. Strain difference in establishment of mouse embryonic stem (ES) cell lines. *Int J Dev Biol* **38**, 385-390 (1994).

160 Auerbach, W. *et al*. Establishment and chimera analysis of 129/SvEv- and C57BL/6-derived mouse embryonic stem cell lines. *Biotechniques* **29**, 1024-1028, 1030, 1032 (2000).

161 Lupton, S. D., Brunton, L. L., Kalberg, V. A. & Overell, R. W. Dominant positive and negative selection using a hygromycin phosphotransferase-thymidine kinase fusion gene. *Mol Cell Biol* **11**, 3374-3378 (1991).

162 Schwartz, F. et al. A dominant positive and negative selectable gene for use in mammalian cells. *Proc Natl Acad Sci U S A* **88**, 10416-10420 (1991).

163 Mansour, S. L., Thomas, K. R. & Capecchi, M. R. Disruption of the proto-oncogene int-2 in mouse embryo-derived stem cells: a general strategy for targeting mutations to non-selectable genes. *Nature* **336**, 348-352 (1988).

164 Wong, E. A. & Capecchi, M. R. Homologous recombination between coinjected DNA sequences peaks in early to mid-S phase. *Mol Cell Biol* **7**, 2294-2295 (1987).

165 Zimmer, A. & Reynolds, K. Gene targeting constructs: effects of vector topology on co-expression efficiency of positive and negative selectable marker genes. *Biochem Biophys Res Commun* **201**, 943-949 (1994).

166 Hoess, R. H. & Abremski, K. Interaction of the bacteriophage P1 recombinase Cre with the recombining site loxP. *Proc Natl Acad Sci U S A* **81**, 1026-1029 (1984).

167 Rank, G. H., Arndt, G. M. & Xiao, W. FLP-FRT mediated intrachromosomal recombination on a tandemly duplicated YEp integrant at the ILV2 locus of chromosome XIII in *Saccharomyces cerevisiae. Curr Genet* **15**, 107-112 (1989).

168 Schwenk, F., Baron, U. & Rajewsky, K. A cre-transgenic mouse strain for the ubiquitous deletion of loxP-flanked gene segments including deletion in germ cells. *Nucleic Acids Res* **23**, 5080-5081 (1995).

169 Schindehutte, J. *et al*. *In vivo* and *in vitro* tissue-specific expression of green fluorescent protein using the cre-lox system in mouse embryonic stem cells. *Stem Cells* **23**, 10-15 (2005).

170 Leneuve, P. et al. Cre-mediated germline mosaicism: a new transgenic mouse for the selective removal of residual markers from tri-lox conditional alleles. *Nucleic Acids Res* **31**, e21 (2003).

171 Berton, T. R. *et al*. Characterization of an inducible, epidermal-specific knockout system: differential expression of lacZ in different Cre reporter mouse strains. *Genesis* **26**, 160-161 (2000).

172 Li, P. *et al*. Germline competent embryonic stem cells derived from rat blastocysts. *Cell* **135**, 1299-1310 (2008).

173 Tong, C., Li, P., Wu, N. L., Yan, Y. & Ying, Q. L. Production of p53 gene knockout rats by homologous recombination in embryonic stem cells. *Nature* **467**, 211-213 (2010).

174 Yang, H. *et al*. Subspecific origin and haplotype diversity in the laboratory mouse. *Nat Genet* **43**, 648-655 (2011).

175 Murray, K. A. & Parker, N. J. Breeding genetically modified rodents: tips for tracking and troubleshooting reproductive performance. *Lab Animal* **34**, 36-41 (2005).

176 Festing, M. F. W. & Peters, A. G. in *The UFAW Handbook on the Care and Management of Laboratory Animals: Terrestrial vertebrates* Vol. 1 (ed Trevor Poole) 28-44 (Blackwell Science, 1999).

177 Silver, L. M. *Mouse Genetics: Concepts and Applications.* 1st edn, (Oxford University Press, 1995).

178 The Jackson Laboratory. in *The Jackson Laboratory Handbook on Genetically Standardized Mice* eds Kevin Flurkey, Joanne M. Currer, E. H. Leiter, & B. A. Witham) (The Jackson Laboratory, 2009).
179 White, W. J. in *Normative Biology, Husbandry, and Models* Vol. 3 *The Mouse in Biomedical Research* (eds J. G. Fox et al.) Ch. 8, (Academic Press, 2006).
180 National Research Council. *Guidelines for the Care and Use of Mammals in Neuroscience and Behavioral Research*. (National Academies Press, 2003).
181 Castelhano-Carlos, M. J., Sousa, N., Ohl, F. & Baumans, V. Identification methods in newborn C57BL/6 mice: a developmental and behavioural evaluation. *Lab Anim*, 88-103 (2009).
182 Schaefer, D. C., Asner, I. N., Seifert, B., Burki, K. & Cinelli, P. Analysis of physiological and behavioural parameters in mice after toe clipping as newborns. *Lab Anim* **44**, 7-13 (2010).
183 Cover, C. E., Keenan, C. M. & Bettinger, G. E. Ear-tag induced *Staphylococcus* infection in mice. *Lab Anim* **23**, 229-233 (1989).
184 Everitt, J. I. *et al*. Metal ear tag-induced foreign body tumorigenesis in p53+/- mice. in *AALAS 53rd National Meeting*, **87** (AALAS, 2002).
185 Baron, B. W., Langan, G., Huo, D., Baron, J. M. & Montag, A. Squamous cell carcinomas of the skin at ear tag sites in aged FVB/N mice. *Comp Med* **55**, 231-235 (2005).
186 Fitzgerald, J. et al. Evidence for articular cartilage regeneration in MRL/MpJ mice. *Osteoarthritis Cartilage* **16**, 1319-1326 (2008).
187 Han, M., Yang, X., Lee, J., Allan, C. H. & Muneoka, K. Development and regeneration of the neonatal digit tip in mice. *Dev Biol* **315**, 125-135 (2008).
188 Le Calvez, S., Perron-Lepage, M. F. & Burnett, R. Subcutaneous microchip-associated tumours in B6C3F1 mice: a retrospective study to attempt to determine their histogenesis. *Exp Toxicol Pathol* **57**, 255-265 (2006).
189 Tillmann, T. *et al*. Subcutaneous soft tissue tumours at the site of implanted microchips in mice. *Exp Toxicol Pathol* **49**, 197-200 (1997).
190 Engel, E. *et al*. Tattooing of skin results in transportation and light-induced decomposition of tattoo pigments--a first quantification in vivo using a mouse model. *Exp Dermatol* **19**, 54-60 (2010).
191 Gopee, N. V. *et al*. Response of mouse skin to tattooing: use of SKH-1 mice as a surrogate model for human tattooing. *Toxicol Appl Pharmacol* **209**, 145-158 (2005).
192 Pritchett-Corning, K. R. *et al*. *Handbook of Clinical Signs in Rodents and Rabbits*. 2nd edn, (Charles River, 2011).
193 Brayton, C. in *Diseases* Vol. 2 *The Mouse in Biomedical Research* (eds J. Fox et al.) Ch. 25, 623-717 (Academic Press, 2007).
194 Smith, R. S., Roderick, T. H. & Sundberg, J. P. Microphthalmia and associated abnormalities in inbred black mice. *Laboratory Animal Science* **44**, 551-560 (1994).
195 Malocclusion in the laboratory mouse. *JaxNotes* **489** (2003).
196 Hydrocephalus in laboratory mice. *JaxNotes* **490** (2003).

197 Korenaga, T. *et al*. Tissue distribution, biochemical properties, and transmission of mouse type A AApoAII amyloid fibrils. *Am J Pathol* **164**, 1597-1606 (2004).

198 Wahlsten, D. Deficiency of corpus callosum varies with strain and supplier of the mice. *Brain Res* **239**, 329-347 (1982).

199 Meng, H. *et al*. Identification of *Abcc6* as the major causal gene for dystrophic cardiac calcification in mice through integrative genomics. *Proc Natl Acad Sci U S A* **104**, 4530-4535 (2007).

200 Poltorak, A. *et al*. Defective LPS signaling in C3H/HeJ and C57BL/10ScCr mice: mutations in Tlr4 gene. *Science* **282**, 2085-2088 (1998).

201 Simpson, E. M. *et al*. Genetic variation among 129 substrains and its importance for targeted mutagenesis in mice. *Nature Genetics* **16**, 19-27 (1997).

202 Clapcote, S. J. & Roder, J. C. Deletion polymorphism of *Disc1* is common to all 129 mouse substrains: implications for gene-targeting studies of brain function. *Genetics* **173**, 2407-2410 (2006).

203 Taddesse-Heath, L. *et al*. Lymphomas and high-level expression of murine leukemia viruses in CFW mice. *J, Virol*, **74**, 6832-6837 (2000).

204 Bowman, J. C. & Falconer, D. S. Inbreeding depression and heterosis of litter size in mice. *Genet Res Camb* **1**, 262-274 (1960).

205 Pritchett-Corning, K. R., Shek, W. R., Henderson, K. S. & Clifford, C. B. *Companion Guide to Rodent Health Surveillance for Research Facilities*. 2nd edn, (Charles River, 2010).

206 Osada, T. et al. Male reproductive defects caused by puromycin-sensitive aminopeptidase deficiency in mice. *Mol Endocrinol* **15**, 960-971 (2001).

207 Nagai, M. et al. Mice lacking Ran binding protein 1 are viable and show male infertility. *FEBS Lett* **585**, 791-796 (2011).

208 Boden, M. J., Varcoe, T. J., Voultsios, A. & Kennaway, D. J. Reproductive biology of female *Bmal1* null mice. *Reproduction* **139**, 1077-1090 (2010).

209 Nishimori, K. *et al.* Oxytocin is required for nursing but is not essential for parturition or reproductive behavior. *Proc Natl Acad Sci U S A* **93**, 11699-11704 (1996).

210 Mitchell, M., Bakos, H. W. & Lane, M. Paternal diet-induced obesity impairs embryo development and implantation in the mouse. *Fertil Steril* **95**, 1349-1353 (2011).

211 Bakos, H. W., Mitchell, M., Setchell, B. P. & Lane, M. The effect of paternal diet-induced obesity on sperm function and fertilization in a mouse model. *Int J Androl* (2010).

212 Robker, R. L., Wu, L. L. & Yang, X. Inflammatory pathways linking obesity and ovarian dysfunction. *J Reprod Immunol* **88**, 142-148 (2011).

213 Hoag, W. G. & Dickie, M. M. A comparison of five commercial diets in two inbred strains of mice. *Proc Anim Care Panel* **10**, 109-116 (1960).

214 Smith, B. K., Andrews, P. K. & West, D. B. Macronutrient diet selection in thirteen mouse strains. Am J Physiol Regul Integr Comp Physiol 278, R797-805 (2000).

215 Teter, B. B., Sampugna, J. & Keeney, M. Milk fat depression in C57Bl/6J mice consuming partially hydrogenated fat. *J Nutr* **120**, 818-824 (1990).
216 Van der Meer, E., Van Loo, P. L. & Baumans, V. Short-term effects of a disturbed light-dark cycle and environmental enrichment on aggression and stress-related parameters in male mice. *Lab Anim* **38**, 376-383 (2004).
217 Browman, L. G. Light in its relation to activity and estrus rhythms in the albino rat. *Journal of Experimental Zoology* **75**, 375-388 (1937).
218 Bronson, F. H. Light intensity and reproduction in wild and domestic house mice. *Biol Reprod* **21**, 235-239 (1979).
219 Rasmussen, S., Glickman, G., Norinsky, R., Quimby, F. W. & Tolwani, R. J. Construction noise decreases reproductive efficiency in mice. *J Am Assoc Lab Anim Sci* **48**, 363-370 (2009).
220 Reynolds, R. P., Kinard, W. L., Degraff, J. J., Leverage, N. & Norton, J. N. Noise in a laboratory animal facility from the human and mouse perspectives. *J Am Assoc Lab Anim Sci* **49**, 592-597 (2010).
221 Hunt, P. Lab disinfectant harms mouse fertility. Patricia Hunt interviewed by Brendan Maher. *Nature* **453**, 964 (2008).
222 Drickamer, L. C. Seasonal variation in litter size, bodyweight and sexual maturation in juvenile female house mice (*Mus musculus*). *Lab Anim* **11**, 159-162 (1977).
223 Lee, T. M. & McClintock, M. K. Female rats in a laboratory display seasonal variation in fecundity. *J Reprod Fertil* **77**, 51-95 (1986).
224 Miyamoto, H. & Chang, M. C. *In vitro* fertilization of rat eggs. *Nature* **241**, 50-52 (1973).
225 Whittingham, D. G. Fertilization of mouse eggs *in vitro*. *Nature* **220**, 592-593 (1968).
226 Kimura, Y. & Yanagimachi, R. Intracytoplasmic sperm injection in the mouse. *Biol Reprod* **52**, 709-720 (1995).
227 Dozortsev, D., Wakaiama, T., Ermilov, A. & Yanagimachi, R. Intracytoplasmic sperm injection in the rat. *Zygote* **6**, 143-147 (1998).
228 Tsang, W. H. & Chow, K. L. Cryopreservation of mammalian embryos: Advancement of putting life on hold. *Birth Defects Res C Embryo Today* -**90**, 163-175 (2010).